MILESTONES
IN HUMAN
EVOLUTION

MILESTONES IN HUMAN EVOLUTION

Edited by
Alan J. Almquist
Anne Manyak

WAVELAND

PRESS, INC.
Prospect Heights, Illinois

For information about this book, write or call:

Waveland Press, Inc.
P.O. Box 400
Prospect Heights, Illinois 60070
(708) 634-0081

Dedication

This volume is dedicated to

RUTH MOORE

science writer, humanitarian, and friend

Ruth Moore received her BA and MA degrees from Washington University and went on to have a remarkable career in journalism, reporting and writing. She worked as a reporter for the *St. Louis Star Times*, as a correspondent for the *Chicago Sun*, and as an assistant editor for *Kiplinger Magazine*. Although she was formally trained in economics and political science, she developed her talents most notably as a science writer.

Her first major publication, *Man, Time, and Fossils* (1953), is an account of the evolution of humankind. This was not an easy task.

The study of human evolution in the 1950s, like that of today, was multidisciplinary, complex, and fraught with scholarly disagreement. Nonetheless, Moore achieved her goal. *Man, Time, and Fossils* has been translated into ten different languages and has been acclaimed by scholars and critics and the lay public. This can be attested to by the following unpublished letter to Ruth Moore written by Kenneth Oakley.

Dear Miss Moore,

I think it was very kind of you to send me a copy of your book *Man, Time, and Fossils* in the English edition. I have enjoyed reading it immensely, and would like to congratulate you on the skillful way in which you maintain the reader's interest by an exciting style of presentation, while at the same time driving home a great deal of factual information. To my mind, it is by far the most successful book on evolution from the paleontologist's angle that has been written. My colleagues agree with me, and some very excellent reviews of your book in English papers and journals show that it is widely appreciated. I like the way in which you tell the story of fluorine dating.

You will be interested to hear the latest development in this field of mineral-dating. After I had given an account of the fluorine work at a meeting of the Geological Society of London in 1949, Dr. Charles Davidson of the Atomic Energy Division of the Geological Survey, told me that uranium was apparently fixed by mineral phosphates, including fossil bones and teeth. At that time he could not help to get any of our specimens tested, and I had difficulty in getting anyone else to make a uranium determination. However, the uranium content of a specimen can be estimated by counting its radioactivity, and when I was in South Africa last year the Atomic Energy workers there tested some fossil bones for me in this way, but they required each specimen to be ground down to a fine powder.

After publication of our Piltdown results last November, Davidson's interest was reawakened and he offered to test a series of specimens from Piltdown and other sites using a method of counting which did not involve any destruction of material. Each specimen is placed in a "lead castle" with a Geiger counter "window" against the specimen, and the beta radiations per minute are then recorded. When the Piltdown specimens were assayed in this way, the results confirmed those of fluorine analysis, and gave us some valuable new information about the associated fossil elephant teeth.

As a result of a very detailed study of the Piltdown assemblage, we have now concluded that everything recorded from the gravel is fraudulent. Even the beaver bones were artificially stained to match the gravel and had been "planted." We were soon quite sure that the elephant and mastodon teeth had been "planted" because it had always been difficult to explain why they should have occurred in a bunch in gravels which were so much later in age. But whence had they been obtained.

The elephant is a type rarely if ever recorded before in Britain. The molar fragments proved to have a radioactivity of over 200 counts per minute, yet the highest count of any fossil from deposits of the same age (Villafranchian) in Britain is 25! We have tested Villafranchian fossils from a large number of localities in Europe, Asia, and North America. Only one site has yielded to us a specimen with the same radioactivity as the Piltdown elephant — a site in Tunisia. (You will realize that uranium like fluorine only indicates the relative dates of specimens if those which are being compared are all from the same site or area). In a week or two I shall send you some publications on these new results.

With kind regards, and again many thanks.

Yours sincerely,

[signed] Kenneth Oakley

In addition to *Man, Time, and Fossils*, Moore published or collaborated on the publication of six more titles. In 1955, intrigued with information she had collected about Charles Darwin while writing her previous book, Moore wrote his biography entitled *Charles Darwin*. In 1956, Moore published *The Earth We Live On: The Story of Geological Discovery*. This book traces the historic progression of the geological sciences and reports the most up to date developments of the day. *The Coil of Life* (1961) explores the genetic and biological basis of life, while *Niels Bohr: His Life, His Science and the World They Changed* (1966) is a biography of the notable physicist. Moore's final independent work, *Man in the Environment* (1975), reveals her sensitivity to and ethical standing on environmental issues, population growth, and our ultimate responsibility in saving our planet. Written in conjunction with an extensive environmental exhibit at the Field Museum of Natural History of Chicago, this book makes the visible exhibit verbally explicit.

Contents

Preface

The study of human evolution is important as a part of general education. But recent advances are being published in many different and diverse scientific journals, museum bulletins, and other sources which are difficult for the average instructor or serious student to obtain. In 1971, S. L. Washburn founded the Society for the Study of Human Evolution. The sole purpose of the society was to produce a series of volumes entitled *Perspectives on the Study of Human Evolution*. Through a cooperative effort by researchers in a number of institutions, these volumes brought together useful information about recent events and breakthroughs in the field of human evolution.

Continuing in that tradition, *Milestones in Human Evolution* is an up-to-date, accessible assemblage of articles which represent some of the major kinds of evidence useful in understanding the human past. No single volume can do justice to the field, but we hope that this new volume will further help to unravel the complex problems of our evolutionary history.

This compilation of papers honors Ruth Moore and her important contributions to science and education. Her special interests were earth history and human evolution, the latter, of course, being the subject through which Dr. Washburn became associated with her. Ruth Moore was an exceptional writer in that she could take highly specialized and often controversial information and synthesize it in such a manner that the lay person could understand both the issues and relevance of the matters. For this reason, *Milestones in Human Evolution* has been written for undergraduates — readers who have no great depth in the field of biological anthropology, but wish to familiarize themselves with the latest controversies, interpretations, and new data. Additionally, this volume focuses on issues that will equally serve upper division students, graduates, and teachers wishing to review the current state of the field.

The organization of this volume follows the title of Moore's best-known work, *Man, Time, and Fossils: The Story of Evolution* (New York: Alfred A. Knopf, 1953). Her goal in this work was to present the story of human evolution to the general public by "interpret[ing]

the work of the expert . . ." (Moore, 1953: vii). In the 1970s, Moore
and S. L. Washburn collaborated to produce *Ape Into Human: A
Study of Human Evolution* (Boston: Little, Brown & Co., 1980). In
this book, they agreed that one of the most important aspects of
the study of human evolution is that it is dynamic and, therefore,
in need of continual revision. Geared for the undergraduate, *Ape
Into Human* presented the knowledge as it stood in the seventies.
In the tradition of evolutionary studies, much of today's scientific
knowledge is different. *Milestones in Human Evolution* is dedicated
to the spirit of evolutionary studies by presenting what we know
at this instant.

Each of the first three sections in this volume is introduced with
a short essay by S. L. Washburn. As one of the most significant
contributors to the field of human evolutionary studies, Washburn
presents an overall picture of biological anthropology today and of
where he feels it might go in the next decade. In Section I, "Man,"
former students present their various findings on nonhuman
primate behavior, having learned from long-term field studies in
the 1960s that the study of nonhuman primates is an important
source of data for learning about human evolution. The subject of
time in human evolutionary research and its conceptualization has
played a vital role in the interpretation of the fossil remains. The
articles contained in Section II, "Time," aim to show that our
changing concepts of geological time have significantly altered ideas
on human evolution. The articles in Section III, "Fossils," illustrate
how an examination and an appreciation of the fossil and
archaeological record contributes to any understanding of
evolution.

In all of her books, Ruth Moore showed interest in the historical
background leading up to major discoveries and, in addition, where
these discoveries might lead in terms of future research. While the
final section, "Future," was not part of the title of Moore's original
work, we include it here in the spirit of her interests in "What comes
next?" Section IV, therefore, explores the usefulness of evolutionary
studies in general and the possible directions new studies might
take.

AJA
AM

Acknowledgements

This volume was the result of the work of many individuals in the Society for the Study of Human Evolution. First, we would like to thank Professor Sherwood L. Washburn for his many hours of consultation and his contributions of ideas and perspectives on the subjects at hand. The editors would also like to thank Margit Aarons and Sabina Johnson for editorial assistance in the preparation of the manuscripts and in bibliographic research for this volume. We would further like to thank Norman Moore, the executor of his late sister's estate, who administered her bequest to the Society for the Study of Human Evolution, which, in turn, funded the preparation of this volume. Finally, we would like to thank Gordon and Anne Getty, who directly and indirectly have contributed to numerous paleoanthropological investigations and made possible much of the research on which the collaborators of this volume have reported.

Introduction

S. L. Washburn

The study of human evolution has changed from an old-fashioned, largely speculative effort into a modern science. Fifty or sixty years ago, the fossils were few, the theories contradictory, and the record complicated. Because of today's numerous fossils, accurate methods of determining age, biochemical techniques for measuring relationships, and new ways of interpreting the information, progress no longer follows the logical pattern of scientific development that was taught many years ago. As more fossils were found and behavior studied, the facts often turned out to bear little relation to the speculations. In retrospect, it can be seen that unexpected changes came primarily in the areas where there had been the most certainty. The same fate may lie in store for many of the ideas we hold today.

Our ancestors, like other mammals, had to act. They had to act on the basis of conclusions they made about the world around them, some of which were wrong or misleading. Before the development of science and the techniques for refining the information fed to the brain, the world of our ancestors appeared small and flat and was understood in terms of subjective perception and experience. Magic, spirits, ghosts, and monsters, can be, and often are, a part of this consciousness. The brain which believed that the elements were only earth, air, fire, and water is the same brain that today masters modern physics and chemistry. The study of human evolution is simply one case in the ongoing process of freeing human thought from the inevitable mistakes and limitations of subjective perception alone. Viewed in this way, the study of human evolution is a critical liberating process. The problems in the study of our history come primarily from fundamental misunderstandings of what the brain is able to do, from confusing the power of thought with the power of techniques and from confusing the fun of emotional cerebration with the stark limitations of rational thought.

From an evolutionary point of view the human brain is both a remarkable problem-solving mechanism and a problem-creating device. Although, in retrospect, these two functions may seem easily separable, this is not the actual case. It is by no means easy to separate rational beliefs from what the future will regard as gross superstition. Our ancestors, who relied on subjective perception and experience as a means of attaining information, felt it was obvious that the earth is flat, that the sun goes around it, that there is spontaneous generation, and that one's own customs are right. This is the rational-emotional common-sense truth shared by humankind. It is equally certain and embedded in custom and belief that spirits and magic exist and that these may be controlled, at least to some extent, by both special customs and talented people. The brain is a remarkable organ, remarkable for its adaptive ability, intelligence, and power of analysis. It is also remarkable in its talent for giving misleading answers and valuing the mistakes as much as positive solutions.

This is well illustrated in the history of the study of human evolution. First, the possibility of evolution was denied. Later, humans were categorized as distantly away as possible from other primates. Then the possibility of immunological classification, of continental drift, of small-brained tool makers, of radiometric dating, and of determination of dates from molecular comparisons *all* were denied, often before the rapidly accumulating evidence could even be usefully presented.

Controversies over theories of human evolution are only a special case of the far-wider questions of the meaning of rationalization and the nature of thought. The evolutionary answer to these questions is clear. The brain never evolved as an adaptation to the problems of the modern world, and the realization of its still-primitive limitations is an essential step in freeing human beings from the past.

The study of human evolution has had a stormy history and has often been contested. Theories of special creation persist in influencing perceptions about the origin of our species to this day. A majority in Western society either deny that human beings are the result of evolution or simply pay no attention to the implications of biological science. When feelings are strong, contesting groups tend to be polarized and discussions of human evolution easily degenerate into dogmatic claims by opposing parties. As a consequence, scientists have found it extremely difficult to maintain that human beings have evolved and, at the same time, to admit the great uncertainties concerning the time, place, and manner of our evolutionary history.

The purpose of this book is to proffer a perspective on human nature based on evolutionary theory and the many fields which have contributed knowledge to it as follows.

Development in Evolutionary Thought

Since Darwin published the *Descent of Man* in 1871, there have been major advances in evolutionary thinking. The rediscovery of Gregor Mendel's laws in 1900 led to an understanding of the nature of variability and to quantifying genetic change (Bowler, 1989). The study of genetics, the discovery of fossils, and general scientific advances led to a synthesis of evolutionary thinking which was clearly stated by 1940. In this country, Theodosius Dobzhansky, Ernest Mayr, and George Gaylord Simpson were the leaders in bringing the fossil record into agreement with genetics. The history of the principal problems is covered most usefully in Mayr's *Evolution and the Diversity of Life* (1976), and the philosophical issues are summarized by Dobzhansky et al. (1977). More recently, Mayr (1992) has written an excellent encapsulation of Darwin's work and the influences that played upon him in the production of the *Origin of Species* (1859).

Obviously, a thorough treatment of the history of evolutionary theory lies well beyond the scope of this book, but issues in theory that have caused great confusion must be considered before going on to the specific problems of human evolution. Any theory depends partially on facts, but these facts exist invariably in a climate of opinion. As the neurophysiologist Paul MacLean has pointed out,

> No measurement or computation obtained by the hardware of the exact sciences enters our comprehension without undergoing subjective transformation by the software of the brain (1970:337).

For example, the pre-Darwinian (pre-evolutionary) concept of the biological world was that it was static. This idea was exemplified in the work of Carolus Linnaeus (1707–1778). He proposed a very useful, albeit static, hierarchical arrangement of species in his *Systema Naturae* (1758). Now, instead of static order, the biological view is of a world in flux, in a process of constant change. One of the best examples of this comes from the development of the ideas about continental drift.

The Idea of Continental Drift

The stable continents of traditional geology have proved to be far from stationary. The history of the theory of continental drift and its transformation into the study of plate tectonics has been repeatedly told. Anthony Hallam (1973) and J. Tuzo Wilson (1976) have given particularly clear accounts.

The idea of continental drift was not new; it had been advanced by Alfred Wegener in 1912 and, subsequently, had been defended for years by Wegener (1924) and others. However,

> Wegener's hypothesis was not accepted for half a century because it was premature, because it required the rejection of orthodox geological theories established and accepted for over 70 years, and because Wegener, not having been trained as a geologist, was not regarded as a reliable authority (Wilson, 1976:5).

After very briefly discussing the evidence for a vast southern continent (Gondwanaland), Alfred Romer remarked, "most geologists like to have their continents nailed down tightly" (1964:156). Even in the late 1960s, this wise paleontologist was very guarded in his discussions of the possible previous position of continents.

What led to the rapid acceptance of continental drift by the scientific community and the development of the much more fundamental theory of plate tectonics was technical advance. The floors of the oceans were explored, mountain ranges charted, and the rocks of the ocean floors sampled. It had been assumed that the floors of the oceans were, for the most part, smooth and similar, representing the accumulations of the ages; however, exploration revealed dramatic differences and a totally unsuspected morphology. A ridge of mountains had been discovered in the middle of the Atlantic Ocean. Using methods that were developed for the exploration of oil-bearing rocks, cores were removed from the floor of the ocean near the ridge and further away. The ages of these cores were determined through magnetic reversal and potassium-argon dating. It was found that the youngest rocks were closest to the ridge and the oldest were furthest away. This demonstrated that new rock was welling up at the ridge and, furthermore, that this process had been going on for many millions of years. After this first success, the major outlines of the plates and their rates of drift were determined in a remarkably short period of time. In effect, a theory rejected as absurd in the 1950s became generally accepted a decade later. Plate tectonics not only afforded a new view of the geologic history of the earth, but it also revitalized geology and provided the impetus for the new science of geophysics.

Conceptions of Time and Human Phylogeny

The conception of a universe that is billions of years old and of drifting continents is simply contrary to common sense. Likewise, the common-sense view of ourselves, that is, the view of what we see and what we feel in ourselves, is just as far from useful understanding. The person of whom we are self-consciously aware is a very limited and odd sample of our biology. When we look at a human being we do not see DNA or biochemical complexity, but grossly functioning phenotypes. This largely external point of view was extended by anatomical dissection and the study of physiology. Still, no one could see a gene, and its nature was not determined until scientists analyzed the structure of DNA and its role in heredity. Just as in the case of the measurement of time or the determination of the motions of the continental plates, the new reality was far more complex than had been anticipated. For example, conventional estimates were that human beings had some ten to forty thousand genes (King and Jukes, 1969:7), but current estimates, based on knowledge of DNA, range from fifty to one hundred thousand genes (Adams et al., 1991:1651).

As recently as the 1980s, it has been argued that apes and humans have been separated for more than twenty million years (Pilbeam, 1984). A corollary of the early separation was the view that there never had been an ape-like creature in the ancestry of human beings (Straus, 1949). Human bones are very similar to those of the apes and are basic to the motions used in climbing. In addition, monkey bones are similar to those of a wide variety of primates, as well as to those of many other mammals, being basic to quadrupedal locomotion. Granted this information, it was concluded either that ape and man were related or that anatomical similarities were due to parallel evolution. In order to settle the issue, one might draw on information showing that the whole anatomy of the trunk and arms, the proportions of one to the other, the position of the heart, lungs, diaphragm, intestines, rings in the trachea, many of the muscles, and the joints of arm bones are also similar in man and ape. However, all the added information did not, in fact, settle the issue of divergence, and the majority of scientists still regarded the similarities as the result of parallel evolution, rather than of close genetic relationship. On the basis of the same facts, the human hand has been regarded as primitive, proving a very early separation or, alternatively, proving an ape-like stage in human ancestry. The issue is not the facts, but how human beings form conclusions on the basis of the facts.

The great contribution of the molecular determination of relationship—that is, the determination of genetic relatedness between species through studies carried out on DNA—is that the human element is drastically reduced. If molecular phylogeny is accepted, then comparative anatomy can be reevaluated, and we can more objectively determine when conclusions drawn from comparative studies are accurate or not. In the example given above, the similarities of ape and human trunks may be interpreted as the result of a long common ancestry after the ape-human line separated from the line leading to the monkeys. To generalize, anatomy will fit the molecular phylogeny if anatomical functional patterns are compared. It may not fit, however, if great reliance is placed on isolated anatomical items, e.g., a single tooth, bone or muscle, a suture pattern, a proportion, or a listing of unanalyzed items. By settling some controversies, molecular anthropology frees anatomy from the mistakes of the past and opens the way for a far more interesting and effective behavioral anatomy.

The relationship of human beings to the African apes is clearly a case where traditional information (few fossils, emphasis on anatomical difference, subjective feelings) led to the placement of humans in a phylogenetic position not supported by the genetic evidence. The evidence of the similarity of DNA suggest that ape and humans are very closely related and anatomy can easily be interpreted to further support this position.

In several ways, DNA, amino acid sequences, and immunology afford estimates of biological differences fundamentally unlike those afforded by traditional evolutionary studies. Most importantly, the results are consistent regardless of which method is used or in which laboratory the tests are performed. While there are technical disagreements over which is the most appropriate method, the results are remarkably similar. Traditional methods never produced this agreement. The time of the divergence of the human line had been estimated as being anywhere between one to thirty million years. All the molecular studies, however, show humans and the African apes to be far more closely related than either is to any other primate.

Evolution of Behavior

It is extraordinary, by contemporary standards, for there to have been so few people during most of human history. Even as late as the end of the period of hunting and gathering, only a very few thousand years ago, population estimates are from one to over two hundred square miles per human being. An estimate of ten square miles per human is often taken as a rough estimate. On the other hand, in East Africa there are some ten baboons to a square mile and far more than that in wooded country. Two hundred arboreal monkeys per square mile is common in forests, and in Malaysia two species of small leaf-eating monkeys may reach densities of five hundred per square mile. In the same area, estimates for hunting humans are some fifteen square miles per person. Even the large, arboreal orangutan has densities of three to six individuals per square mile. During a major glacial advance when the South China Sea was drained, there may have been as many orangutans as human beings in the world. Twenty thousand years ago, there were more San (Bushman) ancestors in South Africa than there were Europeans, and there were something on the order of one hundred baboons per one San.

In the fossil deposits, the relative frequency of one hundred baboons per hominid is not uncommon. Leaving aside the caves, which may only show the dietary preferences of certain carnivores, there always is an abundance of monkey bones compared to those of the hominids. At Omo in Africa, analysis of several hundred fossils from a single locality suggest that the hominids are about as frequent as carnivores.

Numerically then, human beings were never very remarkable in their overall population until after agriculture. In fact, if numbers are taken as a measure of the success of a species, then human beings were not very successful until recently. Changing the evolutionary focus from praise of human progress to the question, Why so few? reopens some very fundamental issues about the diet of our ancestors. If human ancestors could eat the same food as monkeys and apes, and if tool using as well as hunting existed among these intelligent bipeds, then why were there not more human beings? The species *Homo erectus* did occupy a very large geographical area—from Europe to East Asia to Africa—but so did leopards, as well as many other comparably sized carnivores. Until the last few thousand years, the most successful primates were monkeys. It was they who reached high population densities and evolved many new species.

It has been suggested that a stage of scavenging came before hunting (Shipman, 1983). According to this idea, the ancestral apes were vegetarian like all the contemporary apes. They picked up meat from kills, gradually developed a taste for it, and later, learned to hunt on their own. This is a logical sequence; however, it receives little support from nonhuman primate field studies. Neither chimpanzees nor baboons, both of whom hunt, are interested in the kills of other animals. Many years ago, Irven DeVore and I noticed that baboons walked past kills and showed no interest in them (DeVore and Washburn, 1963; Washburn and DeVore, 1961). On occasion, however, they chased and often caught hares which they did eat. The baboons studied by Shirley Strum (1983) at Gilgil, Kenya, hunted for hares and newborn gazelles, but also refused dead animals when offered. They even showed no interest in a hare of the same species as the ones they killed. This seems very surprising, and it shows how different the actual behaviors may be from those which human beings think logical. Aside from the actual behaviors of chimps and baboons, three aspects of scavenging should be remembered. First, not many kills occur relative to the abundance of a vegetarian primate. In general, there are many vegetarians per square mile and many square miles per carnivore. On most days there would be no kill from which to scavenge. Second, the Old World was well supplied with scavengers; not only does much of the average kill go to the killer, but lions, leopards, hyenas, jackals, and vultures all scavenge. One of the principal functions of the big male lion is protecting the kill from all the other animals eager to eat it. Scavenging is a highly competitive and dangerous business, and therefore, a most unlikely way to gain a taste for meat. Finally, as shown at both Gombe and Gilgil, a small amount of hunting provides far more meat than does scavenging.

Nonhuman primate field work is essential in producing what I have called "the ape's-eye view." From the ape point of view, hunting is more advantageous and rewarding than scavenging, precisely the reverse of what we humans thought. Monkeys and apes are making a wise choice when they prefer to hunt rather than scavenge.

Obviously, these remarks apply only to the earliest phases of the origin of human hunting. After humans had effective weapons and the ability to kill large animals, they might also have become scavengers. There are fragments of bone in the same sites as the earliest stone tools, so some direct support exists for the idea that hunting started early in human evolution, more than three million years ago. Certainly by the time of the earliest deposits in Olduvai, hunting of a sort was well established.

The problems of the numerical success of our ancestors in relation to meat eating, scavenging, or hunting, seem very different when nonhuman primate field studies are taken into account.

The importance of the ape's-eye view may be illustrated in a number of other ways. Monkeys and apes occupy exceedingly small ranges by human standards. Obviously, they can see long distances and they are intelligent, but their daily rounds are highly limited. The most limited are those of the arboreal forms, but even the ground dwellers (chimp, gorilla, baboon, patas, langur) spend their lives in only a few square miles. In contrast, carnivores need much more space. Every predator requires many game animals. If the Australopithecines were hunters, then they must have been occupying areas far larger than do contemporary nonhuman primates.

Large ranging areas change both intellectual and social reality. The large area provides both the advantages and problems of predators: stones, woods, total resources, the kinds of prey available, and the necessity for more learning and for leaving the group at long distances. Hunters would have to know where to come back to, where a home base was. Hence, hunting may be the primary cause of both the division of labor and the concept of home. The pattern of human hunting is fundamentally different from that of carnivores because humans are so slow, need weapons for killing, and because mothers must carry their babies. In contrast, for example, hunting dogs may run up to thirty-five miles an hour, have long canine teeth (for holding, killing) and chopping teeth, and may leave pups unattended (regurgitating meat for them upon return). Hunting often involves some cooperation—many carnivores, as well as baboons and chimps, cooperate to some extent—however, human cooperation is fundamentally different, because it involves planning before the actual hunt and interaction afterward. Both of these activities require some form of precise and varied communication, presumably language. Human cooperation also extends to the sharing of hunted or gathered food between males and females. Reciprocal economic relations are basic to the human group, and these behaviors are really quite different from the ones so far described in the behaviors of the nonhuman primates.

Evolution of Language

All the major aspects of human activity depend on language. Beliefs are transmitted by language, and it would be hard to imagine religious, economic, or political systems without complex linguistic communication. All peoples have folklore and conventional wisdom transmitted by language. We are so embedded in a linguistic environment that it is hard to believe that people without training and experience do not necessarily understand the underlying structure of their own language. It is equally remarkable that the nature of language in general and its evolution are still under debate.

The common-sense understanding of language is that words have meanings and that meaningful discourse consists of putting the words in proper orders. Learning a language then, from this kind of understanding, consists of learning the meanings of the words and their appropriate order (syntax). From this perspective, people described the sounds made by nonhuman primates, other mammals, and birds by recording the sounds and the situations in which they occurred and then by inferring the meanings from the sound-situation correlation. Monkey communicating systems were thought to be composed of two or three dozen sounds. The human system of language was thought to have originated in the mere adding of more sounds to the repertoire of the nonhuman primates. If a few dozen sounds composed the monkey system, then a few hundred might have been the system for early man, and a few thousand for people today. Perhaps, as the brain grew larger and larger over the last few hundreds of thousands of years, it automatically became capable of managing much greater numbers of sounds.

This view, implicit in much research and speculation, turns out to miss the major points of the uniqueness and the nature of the problem of the origin of human language. The use of language tells us nothing about the nature of language, any more than running shows a knowledge of joints and muscles, or driving proves we know how an automobile engine works. To put the problem in more general terms, one theory is that the little black box of the monkey brain evolved into the big black box of the human brain and that this evolution may be understood in terms of behaviors without worrying about the nature of the contents of the box. The opposing theory is that the human system is based on a further-evolved anatomical structure and that the differences can neither be understood or even adequately described without a consideration of the brain.

Probably, the importance of speech for cooperation in hunting has been greatly exaggerated. For example, Hans Kruuk (1972) has described the hunting behavior of hyenas as highly cooperative, yet they, of course, do not have speech. Language (as defined by both speech and the brain) might be useful in planning a hunt and in discussing success and failure, but very little communication is necessary at the time of the actual kill. Killing does not necessarily lead to sharing behaviors; in addition, the behavior of lions (Schaller, 1972) on a kill is certainly very different from that of chimpanzees (Teleki, 1973), which is different from that of hyenas. The point of stressing speech is that it is the form of communication which makes human behavior possible. The communication is between intelligent human beings, and this involves both speech and the brain. The nonhuman primates are intelligent mammals, and their brains seem to operate very much like human brains; however, they do not have the capacity to speak (Ploog, 1970). Memory, learning, appreciation of time and space, cause and effect, delayed response, order, condition, questions, and so forth, all seem to be common to mammalian brains (MacLean, 1970), but the abilities may be developed to a much greater degree in humans.

Several major conclusions have emerged from studies on the brain. First, the way the two sides of the brain function cannot be told from anatomical study of the two hemispheres. (For an overview of right and left brain functioning, see Springer and Deutsch, 1981; Levy, 1982.) By all existing techniques the two seem identical, except for a slightly larger part in a language-controlling area, the planum temporale, in some people (Restak, 1984:241). Clearly, there could be major differences in the functioning of the brain of fossil humans which would not be apparent, even if we had the entire brain, let alone an endocast. For example, the differences between late *Homo erectus* or Neanderthal and ourselves in the speech-controlling part of the brain might be very great, but this would not be revealed in an endocast. As noted earlier, the evolution of brain function must be judged by the archaeological evidence in terms of what the people were capable of doing. It is perfectly possible that the biology of speech as we know it today was the main factor in the origin of anatomically modern humans and, if so, this occurred with no change in the gross size of the brain from earlier ancestral forms. If the functioning of the language-controlling area of the brain cannot be determined from endocasts, then clearly it is not possible at the present time to tell whether some fossil forms spoke or not.

Secondly, both sides of the brain think. The nonspeaking side solves problems, and it solves problems of form and music better than the speaking side. There is still much to be learned about the kinds of thought and how they are managed and interrelated by the brain. Language gives information and information may affect thought, but the evidence of both the nonspeech hemisphere and the behavior of nonhuman animals suggests that thought does not depend on language (Beer, 1982). Experiments on ape gestural communication, to which we will return, show patterns of thought which seem the same as ours.

The third point, which is not new but is clarified by divided hemisphere and electroencephalogram (EEG) experiments, is that the brain is constantly functioning even though we are not conscious of its activity (Hillyard and Bloom, 1982). The human brain not only thinks without language, but it can manage complex symbolic behavior, deal with multiple input, and comprehend time. It may turn out to be, then, that human behavior is unique from that of the nonhuman primates because only we have the capacity to speak about time, symbols, and thoughts.

Returning to the structure of the brain, speech functions are not only found on one side of the brain but there is localization of the function on that side as well. Hearing and vision feed information into a coordinating area. This is connected forward to the motor area by millions of nerve fibers and to the opposite hemisphere by millions more. Damage to the language areas may very clearly show the differing functions (Damasio and Damasio, 1992). A person may be able to read but not speak, speak but not read. The inputs are normal. Different defects relate to the way the brain processes the information, not to the incoming sensory information.

Localization, however, is more like an emphasis than a sharply defined area. Further, the cortex has six layers, two thirds of which are located down in the fissures. The deep, underlying nuclei are essential for brain function, particularly in the connections between the thalamus and cortex. From an evolutionary or comparative point of view, these deep structures have increased in most mammals in the same way as has the cortex. Hence, the brain is an integrated machine with functional emphasis, not a machine built of sharply defined, independent parts.

One part of the human brain perceives patterns of sounds (words and sentences) and associates these patterns with arbitrary meanings. The patterns are composed of short sounds which have no meanings (phonemes), but the person learning the language is not conscious of these small units upon which the entire basis for language is built. The child learns words and sequences of words; likewise, in learning a foreign language, we also learn words and syntax.

The "open" character of human language is made possible by the phonetic code. A few sounds may be combined in an almost infinite number of ways. Even with the need of modern science for thousands of new words no limit has been approached, and anyone can easily conceive of new words. This is not surprising because the sound code has the characteristic of codes in general: a few elements may be arranged in almost limitless ways.

In order for speech to be produced, the brain must be able to appreciate the patterns of sounds (input), associate them (memory), and reproduce them (output). This is a uniquely human system and, so far, no one has succeeded in teaching even our closest relatives (the great apes) to speak. Chimpanzees that have been brought up with human children and have been given every opportunity to learn still do not speak (Hayes and Nissen, 1971). Over the last few years there have been several remarkably successful attempts to teach apes to communicate using gestures (Gardner et al., 1989). These promising studies were started by Allen and Beatrice Gardner at the University of Nevada. Seeing the failures of those who tried to communicate with chimpanzees using speech, they shifted to sign language. Their chimpanzee, Washoe, responded in a remarkable way (Gardner and Gardner, 1969).

Communication by gesture is now being investigated in several laboratories. Basically, two methods are used. Either human beings spend very large amounts of time with a young ape, teaching sign language, or alternatively, signs are presented on a panel and the ape's performance recorded and rewarded by a computer. Each method has its advantages. In the human-ape mode, the ape may make new combinations of gestures or invent signs, but the human beings know the ape so well that subjective factors may be involved in the interpretation. With the machine, originality is eliminated, but the record is objective and free of human error or bias. Both methods are useful, and they supplement, rather than compete, with each other.

These experiments have been extremely successful in showing that human and ape can communicate far more than anyone anticipated. Particularly after the numerous failures of verbal communication with home-raised chimpanzees, it is remarkable how much can be communicated when the mode is shifted from sounds to signs. The communications demonstrate that basic cognitive factors are far more similar in ape and human than had been anticipated. The apes manage symbols, have a high degree of comprehension, and understand questions without trouble (Herman and Morrel-Samuels, 1990). As long as the problem comes in the form of "vision in, gesture out," the apes do remarkably well. As R. Meyers has put the matter,

> The chimpanzee is a highly intelligent animal that can communicate ideas very well with humans provided it is not required to use its vocal apparatus. The chimpanzee has shown a particular cleverness in using its hands to communicate both by means of sign language and by depressing levers (1978:72).

Granted this situation, do apes have language? This question has been repeatedly asked since the first studies by the Gardners and people feel very strongly about the matter. The problem is that the word *language* is not clearly defined. Language may be used loosely in a wide variety of senses. For example, people may speak of the language of the bees or the language of numbers or, according to the dictionary, to any system of communicating ideas by signs, gestures, "or the like." In this dictionary sense, apes clearly do have language. They may be taught to arrange signs in orders that have meaning, and may convey a remarkable amount of information in this way. However, apes cannot speak, and if one looks at the subject matter in a book on linguistics, one will find that human communication involves many complexities that move beyond gestural communication alone.

From an evolutionary point of view, the issue is the origin of the brain-speech mechanisms and their adaptive importance. If the problem is viewed in this way, apes do not have language. They lack the ability to learn speech under natural conditions even when human beings make major efforts over years of time to teach them.

This probably is so because their capacity for sound production is primarily under the control of the limbic system and is limited by the anatomical structure of their vocal apparatus.

To use the word *language* both for the naturally learned human system based on the recognition of patterns of phonemes *and* for the nonhuman gestural system, which can only be elaborated with great human effort, is misleading and confusing. Still, it means a great deal to many people to say that "apes have language." This is because language is a "magic word." It is used specifically to minimize the difference between ape and human, just as another magic word, *symbol*, is used to maximize the difference. The only kind of symbols which are unique to human beings are verbal symbols, or objects that receive their symbolic character from language.

It should be stressed that fundamental behavior patterns, like walking and talking for human beings, develop over a period of years with a minimum of encouragement or reward. Biological maturation is clearly very important for both. The life span of most mammals is much shorter than the length of time it takes for the human motor cortex to fully mature (about twelve years). Structure of the brain, maturation of that structure, and concomitant activity (play and practice), all are correlated in the development of the behaviors, and behaviors which have been important in the evolution of the species are easily learned (Hamburg et al., 1974:6–7). Doing the behavior may be its own reward. Children, for example, enjoy increasingly successful efforts to locomote or talk. The successful development of these complex behaviors appears to take years of near-countless repetitions. Persistence in the activity is guaranteed, so to speak, by the pleasure it gives.

If speech is treated like throwing or walking—as a very important behavior with an anatomical base—the human-nonhuman comparison may then be made on the basis of structure, development, play/pleasure, environment (training, culture), and adult function. At every point human speech is unique and not closely approximated in any other animal.

The emphasis on the importance of the evolution of the brain in the evolution of speech does not mean that there was not concomitant change in the articulating mechanisms. Production of the phonetic code (rapid production of contrasting, short sounds) requires lateralization of the control of the phonetic mechanisms, and this is unique to humans. Further, the sound-production mechanisms must have evolved in a feedback relation with the brain. The paradigm may be clarified by what occurred in experiments on tool making in the orangutan (Wright, 1972). Food was placed in a plastic box and the cover tied so that it could not be removed. The investigator then struck off a chip of stone with another stone, cut the string, and opened the box. An orang was allowed to watch the whole performance. After a small amount of repetition, the orang picked up the stone, struck off a chip, cut the string and opened the box. The experiment reveals that an orang, in spite of the highly specialized hand, is capable of making and using stone tools. The primary block to tool making is the problem of conceiving of the idea in the first place. In the case of humans, once there was a beginning of stone tool making—and this might have happened with a creature with a brain and hand of ape size and proportions—then the success of the new pattern of behavior would have led to modifications in both brain and hand for more skillful and powerful use.

Apes and monkeys do make a variety of short and contrasting sounds (Altmann, 1967; Struhsaker, 1967); however, the problem in the evolution of speech is that they never appreciated the beginning of a sound code. If there had been a start, the success of the more effective system of communication would have led to the evolution of both brain and facilitating vocal mechanisms.

It should be remembered that what the brain recognizes is complex patterns; sentences rather than shorter units. For example, a child and an adult, speaking very different dialects, may understand each other even though the actual sounds are very different. Speech is not dependent on particular sounds or particular contrasts but on the appreciation of patterns. This is why machine translation fails. A primitive system of sound code communication might have used different sounds from those used today, as suggested by P. Lieberman and E. Crelin (1971) for the Neanderthals. Reconstruction of the sounds of early humans is probably impossible because the primary issue is the nervous control of the sound-producing mechanisms. The sounds used in earliest speech need not be the same sounds used today. However, just as in the case of toolmaking by the orang, the critical problem is for the nonhuman animal to conceive of the concept of communicating information at all.

Mynah birds copy sounds far better than any nonhuman primate, but their sounds are only copies, not parts of a phonetic system related to meanings. The degree of imitation shows that the sounds may be made by an anatomical system which is very different from the human system and stresses the importance of the human brain as making human speech possible, rather than the articulatory mechanisms. After the human mind began to attach arbitrary meanings to sound patterns, there would have been selective pressures for adaptation in the articulatory mechanisms to the new adaptive situation. Presumably, in addition to evolution in the central receiving, analyzing, and effecting structures, there would have been increased control of facial muscles and pharyngeal and laryngeal functions. However, even when the larynx has been surgically removed, speech is possible, and such extreme cases show that speech might have been based on different structures or sounds.

All contemporary peoples have language (similarity in cognition and speech), and there is no evidence that any language is more primitive than any other. Language, like bipedal locomotion, has a biological base, which is common to the species, but it also contains a far greater element of learned behavior. Biology determines linguistic forms far less than it determines locomotion. This freedom to learn accounts for both the power of linguistic communication and some of its problems. The extraordinary adaptability of linguistic communication depends on a high degree of learning and a minimum of biological determination. The reason that the search for linguistic universals has been so futile is that the biology of language cannot as yet be defined except in a most general way, and the diversity of learning obscures all but a few very general features.

In summary, our intellectual traditions include an examination of our biological history which has led to great progress in the understanding of times past. The techniques of modern science yield information which was undreamed of only a few years ago. Frequently, new fossils are found, and there is steady progress in the discovery of the ways of life of our ancestors. In spite of the progress, however, interpretation of the fossils remains highly subjective, and much room exists for different opinions of the course of human evolution regarding time, place, and manner. As facts are always embedded in a climate of opinion, history, and social systems, what is needed in the study of human evolution, in addition to more effective techniques, is a new climate of opinion—a new willingness to discuss the issues, to appraise where we are, and to speculate about the future.

I

MAN

> *The universe is not to be narrowed down to the limits of the*
> *understanding, which has been man's practice up to now, but*
> *the understanding must be stretched and enlarged to take in*
> *the image of the universe as it is discovered.*
> —Francis Bacon. *Parasceve* (Aphorism 4).

In several chapters to follow, behavior will be stressed as pertinent to an understanding of our relationship to the great apes. Animals adapt through behavior, and successful behaviors determine the course of evolution. The fossils (representing the remains of successful as well as unsuccessful experiments) are used to try to reconstruct the evolution of behavior. Locomotor patterns, brain size, and dental-facial evolution are quite well represented in the record of the last four million years, but a great deal of important behavior leaves few traces in the bones.

Traditionally, calculated guesses have been made about the origin of social behavior based on comparisons of the habits of contemporary animals, particularly our nearest relatives. In following this line of reasoning, we should realize that even our closest relatives are separated from us by millions of years. They are not living fossils, but animals whose survival has been based on adaptations different from ours. Some may have changed less than we have, but primarily they have evolved in different ways. The behavioral importance of bipedal locomotion, object using, and intelligence in late hominid evolution should not obscure the fact that other primates have their own distinct evolutionary history, and behavioral comparisons necessarily look at the end results of these different adaptations.

Field and experimental studies on the nonhuman primates have yielded a variety of information contributing to an expanded understanding of human evolution. The papers presented in this section discuss new insights from primate research in three areas. Jane Lancaster reviews what has become a central focus in primate studies, the role of the female as a pervasive and integrative force in social dynamics. Three decades ago, field research concentrated on male roles and male dominance. Since that time, our conceptualizations of how females behave has changed dramatically.

Certainly, the study of behavior and its relationship to ecological parameters has undergone a revolution in thinking. Today, life strategy theories dominate the field and socioecology has become a major explanation of why the nonhuman primates behave and interact as they do. Horst Steklis addresses these issues in his review of current biobehavioral research.

Within the past few years numerous papers and edited volumes (Harcourt and deWaal, 1992) have addressed the problem of how the structure of a group remains cohesive in the face of ongoing individual aggressive acts. Phyllis Dolhinow and Mark Taff's paper on conflict resolution represents new research in this growing area of interest.

The Evolutionary Biology of Women

Jane B. Lancaster

Jane Lancaster (University of New Mexico) is well known in the field of human evolution and the evolution of human behavior and language. She has authored, among numerous publications, Primate Behavior and the Emergence of Human Culture *(1975) and, more recently,* School Age Pregnancy and Parenthood: Biosocial Dimensions *(1986), in which she expands upon a number of earlier themes and discusses evolutionary perspectives as they relate to human adolescence and reproduction. Her focus in this chapter is on women in human evolution, a perspective often overlooked in earlier research, and gives balance to the contributions of both sexes in behavioral studies.*

This paper is written from the belief that many very important dimensions of human behavior and biology can best be understood from the perspective of evolutionary biology and human socioecology. This is especially true of human sex roles and reproduction, an area of biology and behavior closely linked to evolutionary measures of fitness. Just as we recognize that humans have evolved distinctive patterns of bipedalism, feeding, tool using, intelligence, and social groupings, so too we see that evolutionary processes have left a mark on human sex differences in behavior and reproductive biology in terms of patterns of growth, reproductive maturation, sexuality, fertility, birth, lactation, and parental investment in children.

Two important concepts emerge from a perspective drawn from human evolutionary biology. The first is the concept of the environment of adaptation. The environment of adaptation simply refers to the conditions under which particular patterns of behavior and biology evolved. This is a crucial concept in thinking about human beings because we know that our own evolutionary history has been very rapid and that there have been major, fundamental changes in the context of human experience in recent history. In some senses humans should be thought of as hunter-gatherers (a lifestyle in which we have spent 99 percent of our history) now living under an incredible variety of conditions in terms of nutrition, disease, life course parameters, social density and organization, and level and distribution of resources. In spite of such a wide range of variation, this perspective drawn from evolutionary biology views the cross-cultural record of human behavior as expressing not random but predictable variation in the statuses and roles played by women.

The second basic concept drawn from evolutionary biology is that the reproductive interests and strategies of males and females are not identical. The biological and behavioral adaptations of male and female can best be understood *not* as fundamentally complemental and linked in nature but as separate sets that serve the reproductive interests of each sex. This chapter focuses on the biosocial adaptations found in women that further their fertility and the successful rearing of children. It emphasizes the view that a species does not evolve as a single unit but rather that the adaptations of the two sexes must be understood independently of each other and not in terms of binary complementarity. This perspective also emphasizes active female involvement in the unfolding of her life history in terms of the timing and distribution of her reproductive effort in the life course, in the decision whether or not to invest in particular offspring, and in the choice of sexual partners and of the social networks she will access for aid in rearing offspring. In other words, female choice and the facultative adjustment of female reproductive strategies are seen as an integral part of the human nature of women, selected not by virtue of being the female half of a reproductive pair but as being an individual woman reproducing in competition with others to rear her offspring successfully.

Reproductive Strategies and Environmental Resources

Adaptation and Origin of the Human Family

A number of authors have used differential frequency of types of human family organization such as nuclear monogamy or polygyny to argue species' adaptations in an evolutionary scenario of the origins of the human family. Alexander and associates (Alexander et al., 1979) pointed to the high frequency of societies that permit polygyny and to the modest degree of sexual dimorphism in human stature as indicators that the human species is adapted to be "mildly" polygynous. Lovejoy (1981) scanned the fossil record of hominid evolution for evidence of ecologically imposed monogamy based on male provisioning of females and young and pointed to the relatively moderate degree of human sexual dimorphism and to human food sharing as evidence that monogamy is the core adaptation of the human line. Foley and Lee (1989) studied the adaptational gap between nonhuman primates, the fossil record of the Plio-Pleistocene, and modern hunter-gatherers and tried to reconstruct the niche in which protohominids must have evolved their biology and social behavior. Others (Alexander and Noonan, 1979; for recent reviews see Hrdy, in press; Steklis and Whiteman, 1989) have tried to establish the absence of estrus in humans as evidence for selection favoring monogamous attachment between mates. There is a hidden assumption in these endeavors that a species-specific hominid adaptation exists that is the most fundamental, natural, and original, and that all other forms of human mating and family are derived, less natural, or default behaviors practiced by individuals who are constrained from expressing the pattern most supported by human biology, psychology, and behavior.

This paper will pursue a different course suggesting that traditional categories used to describe human systems of mating and raising children obscure our vision of the essential features underlying and predicting the wide diversity found in the cross-cultural record. It follows the views of Irons (1979) and Haldane (1956) who identify behavioral differences between human groups as environmentally induced variation in the expression of a basically similar genotype and who see facultative responses to environmental differences as the essential human adaptation to socioecological variation.

Evolutionary Biology, Socioecology and Life History Strategies

The past fifteen years have witnessed major theoretical advances in the evolutionary biology of behavior as well as a wealth of field studies on animal populations. It is clear that all animals acquire resources from the environment to survive and reproduce and that the ways in which these resources are distributed in space and time are critical to animal systems of mating and rearing offspring (Clutton-Brock and Harvey, 1978; Dunbar, 1988; Emlen and Oring, 1977; Wrangham, 1979, 1980). This body of theory and research presents a series of generalizations that can inform an inquiry into the division of labor between males and females and into human family formation patterns. The first of these is the distinction between *mating effort* (any investment that increases fertility at the cost of other fitness components) and *parental investment* (any investment in an offspring that increases the offspring's fitness at a cost to the parent's ability to invest in other offspring) (Trivers, 1972). Each individual approaching reproduction is faced with a series of alternatives for the allocation of resources for which the ultimate payoff will be reproductive fitness. Such life history parameters as the timing of reproduction in the life course, temporal spacing between reproductive acts, the number and quality of offspring produced, and the differential allocation of energy and risk between acquiring mates and raising offspring will be affected by whether resources are scarce or abundant, clumped or distributed, monopolizable or indefensible, and certain or erratically available.

The features of resource distribution in time and space present themselves differently to individual males and females. Most theoreticians begin by analyzing sex differences in access to the resources that members of each sex need to maximize fitness. This basic theory permits comparisons between sexes and between species in mating and reproductive strategies. For the purposes of analyzing human behavioral evolution, the most fundamental contributions were Trivers's (1972) germinal paper on parental investment strategies and sexual selection and the papers by Clutton-Brock and Harvey (1978) and Wrangham (1979) on how individuals map behavioral strategies onto environmental resources. Although both sexes are faced with trade-offs in the allocation of resources between mating effort and parental investment, there are fundamental differences between male and female mammals in their reproductive strategies, with males tending to seek as many fertilizations as possible without paying too high a cost in risk and competition, while females must seek access to resources to raise their fertilized eggs to adulthood (see table). This means that females will map their reproductive strategies onto the distribution of the resources they need to rear offspring, and males will map onto the distribution of females either directly or indirectly by controlling resources that females want. One of the critical modifiers of this basic dichotomy between male and female reproductive strategies is whether or not females require aid from others to rear offspring, and, if so, whether they turn to their mates, to their kin, or to cooperative nonrelatives for such assistance.

Sex Differences in Reproductive Strategies of Male and Female Mammals

Male Strategies	Female Strategies
Trade-off between *mating effort* and parental investment	**Trade-off between mating effort and *parental investment***
• maximizes matings, constrained by: — cost (male-male competition, search time) — opportunity (availability of females)	• bears the burden of bringing the fertilized egg to adulthood
Male parental investment	**Female parental investment**
• value is based on impact of what male can contribute to the fitness of his offspring • affected by degree of confidence in his paternity	• seeks access to resources or assistance for rearing offspring
Tactics	**Tactics**
• seeks matings, opportunistic and nonselective • invests in offspring only when payoff is higher than seeking more matings • seeks paternity confidence	• seeks access to resources for parental investment • selective in mate choice or behaves in ways to confuse paternity issues • allocates resources to offspring in ways to maximize maternal fitness

With regard to the reproductive strategies of female mammals, evolutionary theory predicts that, since females bear the very heavy biological burden of gestation, birth, and lactation, they should inevitably link their reproductive behavior to the availability of resources to carry the fertilized egg to the status of an independent adult capable of reproduction (Clutton-Brock and Harvey, 1978; Hrdy and Williams, 1983; Wrangham, 1980). In comparison to other large-bodied mammals, the higher primates in general give birth to offspring at a later age, have longer gestation periods, produce fewer young with each gestation, have longer periods of lactation, experience longer intervals between successive births, and produce fewer young during the life span of the adult female (Altmann,

1987). The higher primates and especially the great apes and humans produce high quality, highly invested young, reflecting both great cost and great value to the investing parent. Among monkeys and apes the major burden of producing such valuable, costly offspring falls almost completely on the shoulders of adult females. For them, wide spacing between births allows the adult female to support a single nutritionally *dependent* infant at the same time that she can foster and protect a second, nutritionally *independent* juvenile (Lancaster and Lancaster, 1983).

In contrast, the human family represents a specialized and very basic adaptation that greatly extended the investment parents could make in their offspring, especially furthering the survivorship of juveniles (Lancaster and Lancaster, 1983, 1987). The investment necessary to transform a zygote into an adult is especially heavy for humans since human children are large, develop slowly, and in many societies need access to specialized resource bases (such as bride wealth, dowry, a homestead, or regular employment) in order to begin reproduction themselves. Hence, women face an even heavier burden than other female mammals in their need for reproductive resources. In such elemental human behavioral patterns as the division of labor, family formation strategies, and parental investment patterns, women should have been selected as active decision makers in optimizing their access to resources and their ability to produce healthy, competitive offspring.

As evolutionary biologists we might predict that women should have a reproductive biology that is exquisitely attuned to offspring viability and to the availability of resources in both the physical and social environment since human children require the most extensive and expensive pattern of parental investment of any species yet described. This sensitivity is found at both biological and psychological levels. At the biological level Surbey (1987), Stini (1985), Lancaster (1989a) and others have noted mechanisms that assay and monitor fat storage and current nutritional status in comparison to past nutrition for the processes of implantation, gestation, birth and lactation. In analyzing infertility, miscarriage and premature delivery in women, Wasser and Isenberg (1986) note an evolved pattern of "healthy infertility" that allows the individual to assess physiologically and emotionally present conditions relative to past. For example, they note that regardless of level, episodes of acute stress are more powerful than chronic stress in provoking miscarriages since a high level of acute stress is more easily measured against past experience. Similarly, acute weight loss is more suppressive of fertility than low amounts of body fat *per se*. Such assessments include unconscious and conscious response to shifts in an individual woman's access to environmental resources and social support systems for the establishment and maintenance of pregnancy and lactation. These abilities essentially provide biological and psychological bases for a woman to cut her losses at any point along the line if the resource base necessary to support reproduction suddenly collapses.

The Division of Labor, the Sexual Contract, and Human Parental Investment

The evolutionary history of human beings differentiates humans from their close relatives in terms of both reproductive and parental investment patterns. Among many species females need assistance to rear young successfully. In nonhuman primates this assistance is usually garnered from the female's kin (Wrangham, 1980). Among humans it is most often, but not always, received from a male sexual partner (Irons, 1983; 1988). In comparing human and nonhuman primates, a striking feature of the human adaptation is the commitment of adults to provision weaned offspring during their juvenile phase of development, a period which is so high in risk in other species that it constitutes a selection funnel into which many enter and few survive (Lancaster and Lancaster, 1987). The riskiness of this period rests on the fact that juveniles are, by definition, small, weak, immature, inexperienced, and poor social competitors. Humans, then, are a species in which the prolonged development of young during the juvenile period and beyond demands a major adaptive commitment to parental investment of such magnitude that males often trade off the value of mating effort for parental investment in offspring.

Although the exact evolutionary history of the division of labor is still under dispute in terms of context, timing, and phylogenetic status, there is no disagreement that sometime in human history males and females differentiated their food-getting behavior in ways that allowed males to specialize in accessing resources from high on the food chain—an endeavor that was both unpredictable and dangerous (Lancaster and Lancaster, 1983). In contrast, females specialized in acquiring high-quality plant foods (the reproductive and energy storage organs of plants) as well as animal fats and proteins in small-sized packets—endeavors that are more predictable and relatively low in risk. This specialization of the two sexes in acquiring energy from different levels of the food chain proved to be of great advantage in that female gathering in the short run could underwrite the unpredictable nature of male hunting.

This underwriting of risk not only benefited both male and female adults but it also was associated with another basic human behavior pattern — the feeding of juveniles; that is, offspring who are weaned but not yet reproductively mature (Lancaster, 1989a). Humans are the only species in which not only nursing infants, but juveniles as well, are freed from the responsibility of feeding themselves. Nonhuman primates do not have nutritionally dependent juveniles — at best a monkey or ape mother will permit her juvenile to share her feeding territory and social status or tolerate its scrounging nearby while she feeds. The nutritional independence of juvenile nonhuman primates and group-hunting carnivores comes at a high price — between 70 percent and 90 percent of those born never reach adulthood largely because of malnutrition and attendant diseases following weaning (Lancaster and Lancaster, l983). In contrast, among humans, regardless of the simplicity of their economy, on average 50 percent or more of children born reach reproductive age. The evolutionary success of this behavioral complex of the division of labor, food sharing, and the feeding of juveniles is testified to by the doubling of the juvenile phase of the human life course to ten years compared to approximately a five-year period of juvenility typical of the great apes (Altmann, 1987). In contrast, the length of infancy is similar in the great apes and human hunter-gatherers: about three to four years is the usual period of lactation and birth-spacing interval between two surviving siblings.

In the course of human evolution the postponement of maturation to full adulthood and the lengthening of the period of offspring dependency occurred because of a doubling of the juvenile phase, the most perilous point of the life course for group-hunting carnivores and nonhuman primates (Lancaster and Lancaster, 1983). Human male and female collaboration in the feeding of juveniles has led to a *minimal* profile of the human family that includes at least one male, one female, a nursing infant, and a series of juveniles at different stages of development but all nutritionally dependent on adults. The division of labor and the collaborative feeding of juveniles by adults is a uniquely human evolutionary package that is supported by the role often played by the male as husband-father. It essentially permits a youngster to linger in what for mammals are the perilous juvenile years while it develops the social and foraging skills to function as a highly competitive adult (Johnston, 1982). The expansion of the juvenile phase of the life course during human evolution was bought at the price of greatly increased investment in offspring from both parents or by other related adults.

Numerous authors have tried to link the division of labor between male hunting and female gathering, the feeding of juveniles, and certain changes in the pattern of female sexuality (Steklis and

Whiteman, 1989). Past scenarios for the evolution of the human family and patterns of human female sexuality focus on a phenomenon that has been variously labelled as loss of estrus, continuous female receptivity, or concealed ovulation (Hrdy, in press). Such labels imply the existence of an ancestress, either ape or protohuman, who limited her sexuality to the days around ovulation, refused males during parts of the cycle when she was not ovulating and during gestation and lactation, and advertized impending ovulation with either prominent sexual swellings or other attractive behavioral or pheromonal displays. Such scenarios focus on the hiding of ovulation as part of a package in the evolution of the human family, either as a concession to reduce the cost of mate guarding for a male willing to give aid to a female in the rearing of her offspring, as an out-and-out meat for sex trade, or as a devious ploy to keep a male monogamously mated because he has to attend a female during most of her reproductive cycle since he cannot detect when she might ovulate.

As diverse as these scenarios are, they all depend on two fundamental beliefs: that humans are evolved to form monogamous unions and that, in order to do so, women had to lose estrus so as to hide the timing of ovulation. In spite of the obvious variability in the cross-cultural record in human family-formation patterns (including polygyny, monogamy, and polyandry, and the generally accepted preference for polygyny among most non-Western societies), such scenarios were reasonably plausible before the in-depth study of primates in their natural habitats. Now it is clear that the baboon-chimpanzee pattern of female sexuality with unambiguous advertising of sexual receptivity and impending ovulation is a specialized adaptation associated with multi-male breeding systems. The other apes (gibbon, siamang, gorilla, and orangutan) do not have this pattern of sexuality, and in all probability the chimpanzee pattern is secondarily evolved. Among the higher primates, estrus behavior and assertive female libido tend to be associated with the female having a number of male sexual partners around a single ovulation. All of these males are likely later to be tolerant or even solicitous of the infant conceived. In surveying the specific anatomical and behavioral qualities of human sexuality, Hrdy (in press) characterizes humans as mildly sexually dimorphic in stature and muscularity (about 5 to 12 percent) and as having small testes and a large penis in the male, and sexual receptivity in the female that is situationally dependent. Viewed against a backdrop of higher primate patterns (Hrdy and Whitten, 1987), this complex suggests early human ancestors who lived in one-male, mildly polygynous groups in which a male did not have to continuously engage in male-male competition (either in terms of physical prowess or in volume of sperm), but in which a female might have to compete with other females for male support and attention. The story of the evolution of the human family can no longer be seen as simply hinging on the establishment of a monogamous pair bond based on meat sharing and the loss of estrus.

Lactation, Women's Work, and the
Costs of Multiple Dependents

In spite of the fact that the evolution of the human family has committed males to high levels of male parental investment, especially in the collaborative feeding of juveniles, human females still bear very heavy costs in the care and feeding of their infants. In fact, the single most significant biological reality for most women during most of human history and during the course of most of their adult lives may have been the biology of lactation (Anderson, 1983; Harrell, 1981; Huss-Ashmore, 1980; Lancaster, 1985). In traditional societies where marriage is universal, where reproducing women rarely contracept, and where spacing between births is maintained by lactational infertility (Short, 1987), lactation should be recognized as a nearly continuous biological state of adult women until after menopause.

Until the onset of high fat and high sugar diets associated with modernization, a major challenge to women has been how to maintain adequate levels of lactation virtually continuously over their reproductive careers without draining caloric and nutrient reserves and depleting their health and their ability to work productively in the feeding of juvenile offspring (Prentice and Whitehead, 1987). This need for nutrients and energy to support lactation as well as the feeding of weaned juveniles is exacerbated by the unusual growth pattern of the human brain (Dobbing, 1974; Sacher and Staffeldt, 1974). Unlike those of our close relatives, human infants are born with brains less than 25 percent of their adult size. Most mammals with single births have more than 75 percent of adult brain size at birth. This means that the human brain, which is clearly our most important organ of adaptation and which grows slowly over the four years following birth, is at great risk to insult from fluctuating environments. Human females have evolved a very complex and specialized program of growth, fat deposition, and fertility establishment that helps to shield the human infant from environmental perturbations in food supply. Access to and storage of energy resources is so important to a woman's reproduction that it interacts with the timing of completed growth, menarche, and the establishment of regular ovulatory cycles, as well as the development of the fetus. In fact, energy storage is so significant in women that sexual dimorphism in fat location and amount is greater than dimorphism in stature and muscularity (Hall, 1985; Stini, 1985).

The significance of lactation to the reproductive life history of women is underscored by Brown's (1970) classic analysis of the division of labor in the cross-cultural record. She found that the underlying characteristics of the tasks assumed by women are their compatibility with lactation — they must be low in risk, performed relatively close to home, easily interrupted, and must not require rapt concentration. No other differentials between the sexes (such as variations in strength, stamina, emotional characteristics, or cognitive aptitudes) appear significant in task allocation revealed in the cross-cultural record. Child-care responsibilities have often produced major constraints on what women do in human society (see, for example, Hurtado, Hill and Kaplan, 1992; Lee, 1979), not because of inherent sex differences in aptitude as much as because of the basic mammalian sexual dimorphism in the ability to lactate. Trends associated with modernization, such as reduction in the percentage of women who reproduce, in the numbers of children born to each woman, and in length of breast-feeding and access to breast-feeding substitutes, have reduced and will continue to reduce the differences between men and women in the tasks they assume and the roles they play in human society.

From the perspective of women, the dependence of juveniles was also bought at an especially high price in terms of female autonomy. Compared to their cousins (female monkeys and apes), women find themselves at a special disadvantage due to their dependence on male help in rearing offspring. In humans and a few other mammalian species in which the male parental role is critical to the survival of young, the price that females pay for gaining the economic cooperation and social protection of their mates is male sexual jealousy and male demands for confidence in their paternity of offspring they help to rear (Daly and Wilson, 1982, 1988; Gaulin and Schlegel, 1980). Paternity confidence is a probabilistic concept. Under normal conditions, there is no such thing as 100 percent paternity certainty. A male mammal will have specific degrees of confidence as to his paternity of a particular offspring. Whether or not a given degree of confidence will elicit paternal investment is again probabilistic, depending on the male's other options for investment such as mating effort to seek new opportunities to mate and produce offspring or parental investment in other offspring for whom the male has a higher degree of paternity confidence (Hill and Kaplan, 1988). For example, in a current study by Hillard Kaplan and myself on parental investment by men in Albuquerque (Lancaster and Kaplan, 1991), we found that fully 21 percent of a random sample of over 2,500 men helped to raise at least one child they knew was not their own, a seemingly highly altruistic pattern of behavior for human males. Further analysis, however, revealed that 72 percent of this investment to nonoffspring was directed to the children or kin of the man's reproductive partner. A further 15 percent went to kin of the man, himself. In other words, men in our sample directed their paternal investment to nonoffspring mainly along paths that could be labeled as either mating effort (dependents of his reproductive partner) or kin selection (dependents of his relatives). Only 13 percent (of the 21 percent of men who raised children who were not their own descendants) helped to raise the children of strangers or nonrelatives, clearly a very rare form of "altruism." Although men do invest in nonoffspring, the vast majority of men (86 percent) who invested in any children invested exclusively in their own offspring or raised both their own children as well as some children of other men.

The need for the assistance of other adults in rearing children and the likelihood that the most motivated and dependable "other adults" will be individuals with a genetic interest in the child together put a burden on women that may lead to an exchange of "paternity confidence" for male parental investment. As Draper points out, women are committed to a disproportionate amount of parental work since, unlike males, they cannot recoup one or a few infant or child deaths by simply finding an additional mate (Draper, in press). A women who loses a child has lost a close emotional attachment, but she has also lost irreplaceable reproductive time. A man who loses his entire family may experience an acute sense of personal loss, but he can replace his lost family by establishing one or more additional mating relationships with other women. Draper identifies this fundamental reproductive inequality between the sexes, due to the higher opportunity costs for a woman based on her biological limitations on the number of children she can bear and on the restriction of her fertility to a specific portion of her life course. Draper notes that this higher opportunity cost gives rise to a unique encumberment of the human female for three basic reasons: (1) Children are born in an extremely dependent state and are slow in development. (2) Unlike other female primates who terminate care of the next oldest offspring when a new infant is born, a woman maintains not one but several dependent offspring at a time. With each new child she adds to her encumberments and goes further and further in debt in the sense that her dependents multiply but her own physical reservoir of energy and productivity remains the same. Consequently, (3) in order to rear her offspring a woman must have help. As a result, the human sexual contract (paternity confidence in exchange for protection and economic resources) has consequences for women that reduce their options and behavioral freedom regarding their reproductive life history compared to nonhuman female primates. This reduction in options may often translate into male dominance of their mates, abuse of women, and control of female sexuality (Burgess and Draper, 1989; Smuts, 1992).

Facultative Adjustments in Human Family Formation Strategies

The reproductive biology of women presents itself as an adapted complex of context-sensitive traits at both the biological and the social-behavioral levels (Lancaster, 1986, 1989a). This complex has evolved to support a highly invested, slowly developing offspring that is nutritionally dependent on adults for many years and born into a sibship of multiple young of different ages, needs, and capacities. Human parents must make decisions at many points about how to invest in offspring over the course of a long maturation during which social and physical resources may fluctuate radically (Lancaster and Lancaster, 1987; Trivers, 1972; Wasser and Isenberg, 1986). Just as the reproductive biology of development and the timing of menarche in women has evolved to make phenotypic shifts according to resource availability, so too have human family formation and parental investment strategies (Lancaster, 1986). There is no single evolved pattern to be found in the human life course or in family formation or parental investment strategies. Rather, evolutionary processes have favored humans in a pattern of phenotypic plasticity and behavioral variation. Such a pattern of flexible strategy in the face of environmental and social variability is not unique to humans but is widely distributed in the social evolutionary history of many animal species (West-Eberhard, 1987).

A wide range of variability can be found in human family-formation patterns, including monogamy, polygyny, single-parenthood, and polyandry (Betzig, Mulder and Turke, 1988). These various patterns of family formation and mating have different costs and benefits for participant members; that is, there are separate, nonidentical trade-offs for male and female members of each unit (Hill and Kaplan, 1988; Irons, 1983). These patterns may be viewed as adaptations to variability in social and physical resources both between and within societies. An extremely interesting aspect of cultural variability is the extent to which the underlying reproductive asymmetry of men and women is institutionalized. For example, under circumstances in which fertility is low and monogamy is imposed (either as a result of ecological constraints or social convention), greater coincidence in male and female reproductive strategies is likely. Similarly, in systems in which women depend on their families of origin for help in rearing children rather than on their mates, much more personal and sexual autonomy for women is predicted since they do not have to exchange paternity confidence for paternal support (Irons, 1983; Lancaster, 1989b).

One of the most striking contrasts between reproductive patterns in low-density social systems (hunter-gatherer and horticultural) and high-density, stratified systems is in control over women's sexuality and their seclusion from unrelated males. Traditional stratified systems appear to present extreme institutionalization of the trade-off between paternity confidence and male investment in children. In effect, the families of women compete for access to quality grooms (grooms controlling resources) using a wide variety of paternity certainty practices ranging from female claustration and incapacitation, to virginity tests, extreme penalties for marital infidelity, and high values placed on female modesty, virginity, and chastity.

Boone (1986, 1988) and Dickemann (1979a, 1979b, 1981) document a reproductive strategy affecting daughters in the upper portions of stratified, agrarian societies that includes a complex of cultural traits: (1) dowry competition by the parents of daughters to gain access to quality grooms controlling resources needed to underwrite the daughters' successful reproduction, (2) female infanticide or spinsterhood in upper class families to reduce the number of daughters to be dowered so as not to impoverish the family estate and reduce the ability of their brothers to gain wives, (3) the closeting of women in the home away from stranger males, and (4) a high value placed on bridal virginity and wifely chastity to raise male confidence in paternity for the select grooms who can offer high levels of male parental investment. In the cross-cultural record, societies with such a fixation on control of female sexuality are also ones where there are great differences among men in power and resources, and the intensity with which families protect their daughters' honor is directly correlated with their social status (Dickemann, 1981; Lancaster and Kaplan, in press). Often women of the lower classes have much more personal freedom than high-born ladies who are virtual prisoners of their own and later of their husbands' families.

In contrast to traditional stratified societies in which so much emphasis has gone into the institutionalization of paternity confidence, it is interesting to note that today's world contains proportionally more single-parent, female-headed families than ever before in human history (Lancaster, 1989b). Powerful social and economic forces have led to conditions worldwide where women often do better in raising children without a mate in the household. These patterns are widely found in underclass families in industrialized societies, in once-peasant societies where men now participate in international labor markets such as in the Caribbean and Mediterranean regions, and in developing countries such as Mexico and those in Southeast Asia where women have better employment opportunities in manufacturing than men, perhaps because they present a more docile and exploitable work force. All of these various conditions leading to single-parent, female-headed households, have one element in common — high variability among men in access to resources and a large proportion of men at the bottom of a stratified system who no longer have sufficient and predictable enough access to resources to underwrite reproduction. Women in these conditions look toward their families of origin or to state-supported safety nets for help in rearing youngsters rather than to male mates whose presence in the household may turn out to be more of a liability than a benefit. Under these circumstances women may gain in personal and sexual autonomy in respect to direct male control. This perspective predicts that the rapidly rising rates of single-parenthood worldwide does not occur by default; that is, not because women cannot get men to marry them, but rather because so many men have lost access to reproductive resources. In other words, modern world economies and industrialization create large groups of men who are barred from both the means of production and reproduction. This powerful shift in male and female access to reproductive resources may be one of the most significant changes in modern human history, the long-term implications for human society and human families being as yet unclear.

A second instructive example of the impact of modern changes in the distribution and timing of women's access to resources can be found in the phenomenon of teenage single-parenthood in modern underclasses in industrialized societies. Geronimus (1986, 1987) argues that the vast majority (98 percent) of teen mothers in our society are not inexperienced "babies having babies" but rather women aged fifteen to nineteen, many of whom are underclass, who may well be optimizing the timing of reproduction in the life course in regards to maternal and child health, maternal fertility, and access to resources for the rearing of children. The familiar statistics that suggest the opposite (that delay in first reproduction to the ages of twenty-four to twenty-nine promotes maternal and child health and maternal education and employment) come from comparisons with the life histories of middle class women and not with the appropriate comparison, the teen mothers' sisters and peers who did not reproduce as teenagers. In contrast to most women in this country, the health and fertility of underclass women deteriorates very rapidly in the life course. For them, in fact, maternal fertility and infant survival peak nearly five years earlier than for middle class women (Geronimus, 1987), and access to family help is likely to be greater during the teenage years (Burton, 1990). Furthermore, since in modern societies the cost of raising a child increases with each year of its life, unlike the opposite condition in traditional societies, teen mothers are free to work in their later twenties to underwrite their children's development once they are in school. According to Geronimus, underclass women make the best of a very bad lot by reproducing as single parents while still in their teens.

Concluding Thoughts

A biosocial perspective on women must emphasize the context-dependent nature of both their biological and their behavioral responses to the demands of reproduction. The evolutionary history of human female reproductive strategies has supported phenotypic and behavioral plasticity in ways that optimize a woman's ability to access the resources necessary to produce and rear her children. These include a reproductive biology closely linked to shifts in social and physical resources, permitting an individual woman to adjust her investment at many points along the path of offspring development, and a behavioral pattern of family formation that optimizes women's access to resources through the most likely and predictable social networks. Compared to the reproductive strategies of female nonhuman primates, woman are confronted with a series of adaptations peculiar to our species: (1) a high commitment to the rearing of multiple, nutritionally-dependent young of differing ages, (2) an unending trade-off between two often-conflicting demands of production and reproduction, and (3) a bargain often struck with males for assistance in rearing young in exchange for raising their confidence in paternity.

Modern changes in the distribution of and access to resources necessary for reproduction are having profound impact on human mating and family formation strategies, the relations between the sexes, and parental investment patterns. Although these changes have drawn us far from the original environments of adaptation and may appear novel and even sometimes undesirable, they are, nevertheless, predictable using a perspective founded on evolutionary biology and human socioecology.

Primate Socioecology from the Bottom Up

Horst D. Steklis

Horst Steklis (Rutgers University) is noted for his work in primate ecology. His significant writings include "The Craniocervial Killing Bite: Toward an Ethology of Primate Predatory Behavior" (Journal of Human Evolution, 1978, with G. King) and "Loss of Estrus in Human Evolution: Too Many Answers, Too Few Questions" (Ethology and Sociobiology, 1989, with C. Whiteman). The contribution to this volume reviews the field of primate socioecology, pointing up the value of ecological and demographic research in relation to behavior as a whole.

The order Primates consists of a highly diverse set of nearly two hundred species, including humans. Living nonhuman primate species are widely distributed over the Old and New World, and while most species are found in the tropical biomes of the earth (especially the African and New World rain forests), many also occur in temperate biomes (including grassland, semidesert scrub, deciduous and evergreen forest), and some range into subalpine and alpine biomes (Richard, 1985). Not unexpectedly, primate ecological diversity is matched by diversity in primate biology and behavior: species range in body size from the 60-gram mouse lemur to a 150-kilogram gorilla; diet ranges from specialized insectivory to generalized omnivory; sociality varies from solitariness to gregariousness. Evolutionary biologists believe that there is an order to this potentially bewildering diversity — an order that has been introduced over time through the process of natural selection. This order is reflected in functional associations between the morphology, physiology, behavior and ecology of a species.

Primatologists seek to understand the particulars of these evolved interrelationships between primate biology, ecology and behavior.

Nearly twenty years ago, one of the founders of modern primatology, Sherwood Washburn, suggested that primatology was in a unique, nontraditional position for understanding primates from a holistic, integrative perspective. This was because primatology had its focus on a particular group of animals, the order Primates, rather than on particular aspects of animals, such as their genetics, biology, behavior, or psychology. Washburn (1973a) envisioned the "promise of primatology" to be an integrative discipline, one in which knowledge of the biology of a species (e.g., anatomy, physiology) was firmly linked to facts about its social system and ecology, much as nature had intended it. A significant part of this promise envisioned an enlightened understanding of the evolutionary basis of human behavior and social organization.

We might well ask whether this promise has been fulfilled. I will argue in this chapter that, where the study of primate behavior is concerned, the answer is negative. As I will show, in primatology, as in the field of animal behavior generally, studies of biological bases of behavior (or proximate causation) and evolutionary (or ultimate) causes of behavior are carried out independently. Despite this unfortunate history (and, hence, a "broken promise"), I believe new opportunities exist for exciting and fruitful disciplinary integration — for making amends. The strong development of socioecological theory that has grown out of sociobiology enables us to integrate theory and data with conceptual advances and information gained from the study of proximate mechanisms. Specifically, I will try to connect biological characteristics of individuals with features of social organization. I will propose that characteristics such as behavioral disposition (e.g., temperament or personality and their proximate causes) are shaped by evolution and, thus, hold promise of being linked to variation in features of social organization.

This is necessarily a daring and surely ambitious undertaking, especially within the constraints of a book chapter. In fact, the attempt might rightfully be questioned, given the very complexity of both the physiological bases of behavior and the relation of behavior to social organization and ecology. The task, therefore, requires moving between disciplines like ecology, neurobiology, and animal behavior, and across levels of explanation of behavior, for example, those of physiology, of interacting individuals, and of social structure. Along the way, this effort will raise questions (and probably eyebrows) about the interrelations, and particularly the reduction among levels of explanation. My conviction stems from the rapid development and application of sociobiological theory to primate socioecology. This development has produced a flurry of activity wherein theory has tended to outstrip data and adaptationism has gone rampant. These are the forgivable mistakes of a youthful discipline which should correct themselves as the discipline matures. What is more troublesome, however, are deeper flaws in the enterprise, such as the view that proximate explanations have little to contribute to understanding the evolutionary basis of behavior (Chadwick-Jones, 1987). This view is unfortunate at a time when evolutionary issues point to new and challenging questions about causal mechanisms of behavior. An understanding of these mechanisms can prove to be, in turn, crucial to answering evolutionary questions (Stamps, 1991). Moreover, in recent years, evolutionary theory has provided renewed impetus and fresh theoretical perspectives to physiologists interested in social behavior (Crews and Moore, 1986). This development holds promise for reuniting the long separate study of proximate and ultimate causes of behavior.

Primatology Then and Now
An Overview

While primatology, with its focus on a group of animals, does not in principle limit itself to the study of any one particular aspect of an animal, its practitioners are drawn from several traditional academic departments, each with its own conventions of analysis and interpretation. Unfortunately, these different traditions have historically kept the study of biological mechanisms of behavior separate from the study of behavior in its social and ecological contexts. A sketch of the historic development of primatology makes this clear.

Historical Fragmentation of the Discipline

The post-World War I rise of primatology can be attributed to pioneers in two rather divergent disciplines: psychology (particularly the work of Robert Yerkes) and anatomy (particularly the work of Adolph Schultz). These were soon followed by waves of physical anthropologists, zoologists, geneticists, ecologists, and others. The spread of research interest in primates within the social, life, and medical sciences soon led to the very real worry that fragmentation would undermine the mutual contact so essential for the discipline's progress (Schultz, 1971).

Where the integration of behavior and biology was concerned, primatology's progress was indeed hindered by a perhaps inevitable fragmentation into respective traditional disciplines. Anatomists and geneticists studying primates, by and large, remained interested in issues of taxonomy (the classification of primates). Physiologists became interested in primates as model systems for medical research. They concentrated their efforts on a few hardy, widely available primate species to such a degree that everyone knew "the monkey" referred to in scientific publications was none other than the rhesus macaque (*Macaca mulatta*). The study of physiology and comparative anatomy did not require that research be done in a social setting, much less in the wild; rather, controlled laboratory conditions provided the optimal investigative situation. Thus, the study of primate social behavior and its relationship to ecology was divorced from the exploration of the mechanisms of behavior.

A few researchers were exceptions to this trend. Among these were primatologists and anthropologists, who joined forces with biological psychiatrists to tackle the problem of proximate causation (Kling, Lancaster and Benitone, 1970). Many were Washburn's students who were inspired by his integrative vision of primatology. I suspect the reason that psychiatrists (Bowlby, 1969), rather than physiologists or neuroscientists, first became interested in the biological bases of primate social behavior was their disciplinary interest in unravelling the complexities of human social behavior. During this time, neuroscientists and physiological psychologists focused on issues of clinical relevance — mechanisms of learning, memory, and cognition — without particular concern for the relevance of their findings to social behavior.

This development in primatology paralleled that in the field of animal behavior in general. Also following the development of sociobiological theory (see below), and despite the traditional holistic, integrative aims of ethology (Dawkins, 1989), the study of mechanisms was ignored at the expense of a concentration on the adaptive nature of social behavior. The study of mechanisms was left to a specialized group of ethologists who called themselves "neuroethologists" (Ewert et al., 1983); the study of adaption became the purview of a group of sociobiologists who called themselves socioecologists or, more recently, behavioral ecologists. Dawkins (1989) argues that neither group can claim legitimate research goals apart from those of traditional ethology; however, calling oneself a behavioral ecologist rather than an ethologist appears to be more fashionable. She further asserts that, as a result of this fashionable trend in disciplinary fragmentation, the organisms and behaviors of greatest interest to neuroethologists, such as stereotypical behaviors produced by simple nervous systems (for example, fixed action patterns in invertebrates) had little relevance to the study of complex behavioral strategies that came to occupy the behavioral ecologists.

Primatology Today

In primatology, the consequences of this splintering can be readily illustrated through the predominant emphasis on the adaptiveness of social behavior, attested to in current journals and trade books. Popular primatological textbooks (Jolly, 1985; Smuts et al., 1987) provide excellent coverage of behavioral diversity and ecological adaptation but offer little on the proximate mechanisms of the behaviors described, such as sex, parenting, dominance, or dispersion. Despite the accumulation of data on such mechanisms as the role of hormones in aggression and dominance, this omission persists. This seems all the more surprising, since primatologists have become significantly involved in research on aspects of proximate causation, such as hormones and behavior (Bernstein et al., 1983), and the neurological and neurochemical bases of primate social behavior (Raleigh et al., 1984; Steklis and Kling, 1985).

Nevertheless, both practical and ideological reasons exist for why the results and perspectives of studies on proximate causes of social behavior have not found their way into general discussions of primate behavior. In primatology, few attempts have been made to integrate with the emerging body of socioecological data and theory, in part, because the task seems daunting (Mendoza, 1984; Mendoza et al., 1991, as an exception). Consequently, many primate behavior synthesizers will shrink from the effort without necessarily feeling that something has been left unexplained. More importantly, many will not consider the effort worthwhile because of the antithesis between biological and social levels of explanation, the result of the historical separateness of these lines of investigation. Reducing social behavior to biology thus creates a fundamental problem to which we will return later.

The Development of Socioecology

To appreciate why this is a propitious time for integrating proximate and ultimate causations of behavior, we must look closer at the development of primate socioecology to see how it has led to a renewed interest in the proximate causes of behavior. My review is necessarily selective (for more systematic reviews see Terborgh and Janson, 1986; Richard, 1985).

History of Socioecology

The cornerstones of primate socioecology were laid not by traditional ecologists (who were busy studying plants and small, short-lived animals), but by psychologists and anthropologists who were drawn to nonhuman primates because of their interest in human nature (Richard, 1981). As fieldwork and studies of captive social groups accumulated, a need developed to synthesize ecological and social behavioral data — that is, to synthesize the patterns of the varied forms of primate social organization and their respective habitats with the adaptive nature of social organization. The first and simplest approach was to organize primate societies and ecologies by type, such as uni-male groups, arboreal, frugivorous, and to identify patterns of association between these types. Though perhaps simplistic, this initial typological approach was, in principle, entirely legitimate, given that the majority of primates spend their lives within a social group and that all the important problems of life, including feeding, mating, parenting, and escaping predation, are social problems requiring social solutions. The social group with a particular structure (such as age and sex-class composition) and organization (such as patterns of relationships between group members, including dominance, mating, and rearing systems) can thus be rightly seen as a means of ecological adaptation. Put another way, the social systems of primates (and those of other social species) have been shaped by natural selection in response to ecological contingencies encountered during the history of a species.

Unfortunately, as so often happens with typologies, nature is not easily pigeonholed. While these typologies had heuristic value, the research they stimulated also served to underscore their flaws (Richard, 1981). Important was the realization that species could not fit neatly into types. In some instances of social organization, as much variation exists *within* a species as does *between* species. As a result, the constructed ecological dimensions were overly simplistic: basic categories such as habitat or diet simply could not account for the full range of observed social organizations.

That intraspecific variability in social organization occurs without detectable differences in habitat created particular difficulties for typologies. Possibly, and uncomfortably, it suggests that a primate population might have alternate social solutions to the same ecological problems, or worse, that, at any given point, a group's particular social structure and, hence, its social organization might reflect random demographic changes (Altmann and Altmann, 1979; Rowell, 1979). Thus, variable social structures might result without any particular strategy being more adapted to ecological conditions than another. Such possibilities proved annoying for two reasons. First, one of primatology's close and influential disciplinary neighbors, ethology, was focused on the adaptive nature of species-typical characteristics, such as the classic fixed action pattern. Consequently, ethologists were little interested in individual differences, or within-species variation in general, considering it a nuisance that distracted from their study of species-typical characters (Barlow, 1989). Second, until the early 1970s, prior to the advent of sociobiology, behavioral variation among individuals within groups was not considered evolutionarily significant. Entire social systems or groups, rather than the individual, were treated as the adaptive unit. Both primatologists and ethologists employed group selection as a means of evolutionary explanation (Barlow, 1989).

To explain the function and evolution of social systems, the early 1970s replaced group selectionist theory with individual/gene selection theory (Williams, 1966; Dawkins, 1976). The architects of the new discipline called sociobiology argued that the evolution of sociality and its diverse forms must be explained at the level of reproductive costs and benefits to the individual (Alexander, 1974; Wilson, 1975). In other words, they examined and explained sociality in terms of how behavior contributed to individuals and their relatives' reproductive success. In this view, individual behavior did not evolve to serve the benefit of the group or species; rather, individuals were reproductive competitors. Even acts of altruism were thus explained as genetic selfishness. This theoretical shift refocused attention on individual differences, especially between the sexes (see below). Behavioral variation among members of a species was no longer considered a nuisance, but a phenomenon of potential evolutionary significance (Clark and Ehlinger, 1987). Although sociobiologists, through a focus on the adaptive consequences of individual behavior, sought to reduce behavior to genetic differences among individuals, their effort has not led to the bold genetic determinism so feared by the discipline's critics. Rather, it has led to a conception of animals as relatively labile, though not necessarily cognitive, calculating machines (Barlow, 1989), ones whose behavioral decisions (such as reproductive strategies) are influenced by both past and current social and ecological stimuli.

Explanations of the diverse social organizations between and within species must, therefore, start with a consideration of individual reproductive strategies. The different life histories and energy demands of the two sexes have made us consider female and male reproductive strategies separately. In simple terms, because of the female's disproportionate contribution to offspring, her reproductive success is limited by available nutritional resources; in contrast, a male's reproductive success is limited by the number of females he can inseminate. From this basic difference of the reproductive investment strategies between the sexes (common in mammals), it follows that females will be adapted for efficient harvesting of resources and, thus, more constrained by habitat quality, while male strategy focuses on behaviors that increase their access to female investment (Gaulin and Sailer, 1985). Thus, increases in male body size and fighting equipment, which may be energy costly, will be selected for, even if they decrease foraging efficiency, so long as they increase access to female investment. Since female body size is relatively more constrained by nutritional resources, sexual selection may result in pronounced sexual dimorphism.

Gaulin and Sailer (1985) argue that, if female characteristics (both morphological and behavioral) are indeed more constrained by nutritional resources, then female body weight for a given species should be closer than male body weight to an optimum weight predicted from dietary quality. Such an optimum can be determined on the basis of the Jarman-Bell principle (Gaulin, 1979) which states that related species share a negative relationship between body weight and overall dietary quality. That is, large-bodied primates can subsist on lower quality and generally more abundant foods, like leaves and shoots, because of their lower per-unit-body-weight food requirements; smaller-bodied primates, to keep pace with their higher per-unit-body-weight nutritional requirements, must eat higher quality, but rarer foods, like gums, insects, and small vertebrates (Gaulin, 1979). In a sample of fifty-three primate species, female body weights did indeed deviate less than male body weights from the optimal weights predicted from dietary quality, suggesting that females are "the ecological sex" (Gaulin and Sailer, 1985). Furthermore, Gaulin and Sailer predicted that the extent to which male primates deviate from optimal body weights can be correlated with mating system, or the degree of intermale competition for females. This prediction was confirmed: in species with polygynous or promiscuous mating systems (for example, baboons, where competition for females is high and males show relatively little parental care and high variance in reproductive success), male body weights deviate further from optimality; in monogamous species (such as gibbons, that have less direct male competition for females, less difference in parental investment between the sexes, and little variance in male reproductive success), male body weights do not deviate as extensively. (See, however, Rowell and Chism, 1986, for a discussion of the inherent difficulties in drawing an association between mating system and sexual dimorphism.)

The pressures of differential selection on males and females should also account for behavioral differences. We can predict that female behavior should be attuned to habitat characteristics, like efficient harvesting of nutritional resources, while male behavioral features should be attuned to improved access to females. Female grouping, therefore, should be determined by the distribution and quality of food resources. Male grouping, conversely, should be determined by the distribution of females. In other words, the distribution of males depends on their ability to defend and monopolize groups of females (Emlen and Oring, 1977), which will result in a particular kind of mating system. This hypothesis has been tested by Ims (1988). He found that female spacing in relation to resources does indeed appear to determine male spatial distribution and relationships. Under different ecological circumstances, however, we can expect species' differences in male response to equivalent conditions of female distribution.

Socioecology Today—"Where the Girls Are"

The discussion so far should make apparent that consideration of male and female reproductive strategies has come to play a central role in theories of vertebrate social organization. To summarize the current view: for any population, social structure and organization are the result of the interplay (or compromise) between male and female reproductive interests pursued in particular ecological and demographic circumstances. Further, the behavioral options to carry out these reproductive interests will be limited by evolved life-history characteristics, such as body size (Harvey et al., 1987). Since female spatial distribution depends on habitat, while male distribution is determined by females, the benefits of female social grouping must be looked at to explain variances in social organization.

In primatology, the focus on reproductive costs and benefits has reopened debate on the ultimate causes of sociality (van Schaik, 1983; van Schaik and van Hooff, 1983; Wrangham, 1987; Terborgh and Janson, 1986; Isbell, 1991). The costs of sociality can include: (1) increased food competition with larger group sizes and associated lower reproductive rates; (2) increased disease spread and parasite load; and (3) reduced fecundity due to social stress unrelated to feeding competition (Dunbar, 1989). Three major benefits for females living in multi-female social groups have been proposed: (1) efficient harvesting of food resources; (2) avoidance of predation; and (3) avoidance of infanticide. As described below, considerable disagreement exists concerning which costs and benefits are sufficient or necessary to explain the variation in primate social groupings.

Some primatologists, for example, have downplayed the importance of predation relative to feeding advantages in structuring primate groups (Wrangham, 1980; Rodman, 1988; Isbell, 1991). They argue that group living for females is favored whenever the advantages of cooperative defense of food resources against other groups of females outweigh the costs of intragroup competition. According to Wrangham (1980), the distribution and defensibility of resources (e.g., clumped food, as in patches of fruit trees, compared to evenly dispersed or very large patches of leaves and terrestrial herbaceous vegetation) determine female, and ultimately male, spacing and social relationships. Thus, cooperative defense of food patches by a group of females favors "female-bonded" societies, like those of macaques and baboons, which are characterized by differentiated female relationships (i.e., matrilines and linear dominance hierarchies) and female philopatry (i.e., males emigrate, females remain). In "nonfemale-bonded" societies, such as those of folivorous species like gorillas and some colobine and howler monkey species, female transfer is the rule, and differentiated female relationships should not exist. In this model, then, food distribution alone influences both intra- and intergroup competition among females.

Isbell (1991) has argued that socioecological models must consider the differential effects of food distribution versus food abundance. In her scheme, food distribution determines only the type of intragroup relations, while food abundance determines the nature of intergroup relations. Resource distribution and abundance, she suggests, are sufficient determinants of female relationships within and between groups. Following van Schaik (1989), she distinguishes between two types of feeding competition: "contest competition," involving active aggression over food, and "scramble competition," involving passive behavioral adjustments, like adjusting ranging behavior according to group size. In her analysis of patterns of female aggression and ranging behavior among twenty primate species, she associated clumped distribution of resources with strong, linear female dominance hierarchies and increased day-range length as a function of group size. She linked dispersed resources to absence of female dominance hierarchies and found no relationship between group size and day-range length. The majority of species that are restricted by food abundance showed intergroup aggression and increased home-range size as a function of group size, regardless of type of diet or food distribution. Thus, Isbell suggests that whenever food abundance limits population growth, females will act aggressively toward females of other groups. Further, the distribution of resources (clumped or dispersed) determines the nature of within-group female relationships. However, in those species in Isbell's scheme whose reproductive success seems not to be limited by food abundance and distribution — such as specialized folivores/herbivores, some colobines, howler monkey species, and gelada baboons — we must seek out reasons other than ones related to food to explain within- and between-group female relationships. We might consider predation to have a relatively greater influence on multi-female groups in these species.

Dunbar (1986), for example, has presented strong evidence for the primacy of predation as a cause of multi-female groups in gelada baboons (*Theropithecus gelada*). Social stress, caused by harassment of subordinate females by dominant females, rather than competition for food resources, limits reproductive success in female geladas. Female birth rate is a function of dominance rank and age. Rank, in turn, is inversely associated with the number of anovulatory cycles, suggesting that harassment suppresses ovulation. Since coalitions among females can reduce harassment and its associated costs, much of the competition among gelada is over coalition partners. The abundance and distribution of gelada food (grasses) permit the forming of large groups that provide optimal predator defense; however, gelada females in larger groups suffer relatively greater reproductive costs due to harassment. To avoid these costs, which presumably are higher than the benefits received from predator protection, Dunbar argues, females live in smaller breeding groups consisting of one breeding male and one to ten adult females. The gelada harem group thus appears to be a good compromise between the costs of female harassment and the benefits of avoidance of predation.

In contrast to both Wrangham and Isbell, van Schaik has argued that among diurnal primates, "predation confers the only universal selective advantage of group living" (1983:138). In his view, predation sets the lower limit for group size while feeding competition among group members sets the upper limit. Terborgh independently proposed the same model (Terborgh and Janson, 1986). He argues that resource patch size plays a passive role by permitting such large group sizes as were selected in response to predator pressure. This same study shows that group size, in turn, constrains the social system. Consequently, the very smallest group sizes are associated with the following social systems: solitariness (nocturnal prosimians), monogamy (gibbon and siamang), and polyandry (tamarins and marmosets). Large groups of primates occur when patch size permits, regardless of predation pressure.

A greater diversity of social systems is possible with more group members. Thus, large groups are associated with the following social systems: fission-fusion (chimps and spider monkeys), harem (several colobine and cercopthecine species), or multi-male social systems (baboons and macaques). Van Schaik (1989) has taken this model further by predicting the types of female relationships within and between groups that result from competition for safety and food. He argues that predation pressure forces females to live in cohesive groups, which leads to within-group competition for resources. Depending on the nature of this competition (contest or scramble), groups will show either egalitarian or nepotistic and despotic ranking systems. In species with low vulnerability to predators (the largest arboreal and small nocturnal species), little within-group competition occurs, and between-group competition predominates. Where ecological conditions permit the formation of sizeable groups, these will be characterized by egalitarian social relationships. Van Schaik's predictions have been verified in recent field studies (Mitchell et al., 1991).

The role of predation as an independent cause of variation in primate social groups remains difficult to evaluate directly. This is partly because predatory events tend to be rare and uncontrolled, and partly because little comparative data exists on predation rates (Cheney and Wrangham, 1987). We have had few opportunities, therefore, to test the relative importance of predation in structuring primate groups. In a systematic review of the evidence, Cheney and Wrangham conclude that across primate species "there is a negative correlation between predation pressure and body size . . . there appears to be little relation between predation pressure and terrestriality, degree of sexual dimorphism, group size, mating system, or the number of males per group" (1987:239). The authors conclude that predation has not been the primary cause of the evolution of social structure or morphological traits.

In light of this conclusion, we ought to look in greater detail at some recent field studies that address the potential importance of predation to primate sociality. The first two studies, van Schaik and van Noordwijk (1985) and Isbell et al. (1991), illustrate the difficulty of interpreting the evidence when it is indirect, i.e., correlational. The third study, by Boesch (1991), which includes substantial direct evidence, highlights the problem of across-species and habitat generalization about the particular effects of predation.

The prediction that group sizes should be larger where predation risks are higher was tested by van Schaik and van Noordwijk's comparison of two populations of long-tailed macaques (*Macaca fascicularis*) living in habitats that differed primarily in predation pressure. As predicted, on the island of Simeulue (off the coast of Sumatra), where there are no large felids, groups were smaller than at Ketambe, Sumatra, where the felids occur. However, such correlational evidence from a small sample, while logically compelling, is not conclusive: first, there is no direct evidence of felid predation on these macaques; second, data from another population show that group sizes of long-tailed macaques in Borneo, where there are no large felids, are similar to those at Ketambe (Rodman, 1988).

Isbell et al. evaluated the causal role of predation. They examined data on patterns of vervet monkey (*Cercopithecus aethiops*) group fusions during a period of population decline in order to test enhanced predator detection and avoidance as a cause of group membership. While juvenile vervets appeared as effective as adults in detecting predators, judged by relative frequency of alarm calls given to predators and response to alarm call playbacks, adults alone determined the timing of fusions. Adult vervets remained as a group so long as at least one other adult was present. Since vervet juveniles and adults make similar contributions to predation avoidance, but not to intergroup competition, the authors conclude that the advantage of group living in vervets is more related to resource competition than to predation avoidance. Here, too, the logic is compelling, but the evidence is strictly inferential.

The recent observations presented by Boesch of leopard predation on chimpanzees (*Pan troglodytes*) in the Tai forest, Cote d'Ivoire, are compelling. This study not only provides the first direct evidence of predation on a species previously thought to be free of such pressure, but it also demonstrates the independent effects of predation on social organization and their modulation by habitat. Before completely appreciating the results of Boesch's findings, however, one needs a background on chimpanzee social organization and its presumed causes.

Chimpanzees have a "fission-fusion" type of social organization (Goodall, 1986), wherein a community of chimps splits into foraging parties, often consisting of a single female and her dependent offspring or of related males who also patrol the community borders. Males form long-term bonds with each other, and their social relationships are characterized by dominance and alliances, while relationships among females (the dispersing sex) are relatively undifferentiated and social ties are weak, which is why females usually forage solitarily or with their offspring. Consistent with the resource competition hypothesis discussed earlier, the chimpanzee fission-fusion system is thought to be an adaptation to within-group feeding competition, so intense that high costs are incurred by females foraging in multi-female parties. Males, conversely, form cooperative alliances as a strategy to sequester and defend females against other males (Wrangham, 1986). Along the same lines,

Wrangham has suggested that the closer female-female and female-male social ties of the pygmy chimpanzee (*Pan paniscus*)—as compared to those of the common chimpanzee—and the larger heterosexual feeding parties at Lomako are made possible by reduced within-group feeding competition, due to the relatively greater availability of terrestrial herbaceous vegetation.

With this background in mind, let us now turn to Boesch's study. Over a period of five years, he documented twenty-two interactions between chimps and leopards. Incidents included chimp attacks on leopards, leopard attacks on chimps leading to injury or death, and leopard predation on chimps. These interactions resulted in the death of four chimps and the wounding of six, yielding a minimum estimated predation risk of one chimp being attacked by a leopard about once every three years and of being killed within eighteen years. Boesch suggests that leopard predation is the primary cause of chimpanzee mortality, with leopard predation accounting for at least sixteen to seventeen of forty-eight chimps that disappeared during the five-year period.

Boesch's work clearly underscores the difficulty of making broad generalizations about the relative influence of predation pressure on primate groups. Large body size, for example, does not ensure relative immunity from predation, nor do larger groups necessarily enjoy relatively greater safety from predators. Hypotheses about the influence of predators on social groups will have to be species and habitat specific, taking into account not only the habits of particular predators, but also the prey animal's potential behavioral (and cognitive) defense repertoire.

In addition to protection from predation, females may live in groups to reduce infanticide, a common male reproductive strategy among mammals (Hausfater and Hrdy, 1984). Since this threat to females is posed by unfamiliar males, females can reduce this risk by forming a long-term breeding relationship with one male, resulting in either a monogamous or polygynous mating system. Van Schaik and Dunbar (1990) have proposed two types of monogamy in primates that have evolved in response to different selection pressures. The first type involves a high degree of male parental care, as in the small-bodied callitrichid species, in which both the female and the male benefit from high male involvement in parental care. In the smallest species this may lead to facultative polyandry. The second type, found among larger-bodied primates (gibbons), involves little or no direct paternal care, and some species show facultative polygyny. Many traditional hypotheses advanced to explain this type of monogamy invoke food distribution and joint territorial defense of food resources as causes (Barlow, 1988). Van Schaik and Dunbar (1990) evaluate these proposals and conclude that monogamy that includes active pair bonding among large-bodied primates is best explained by the service the male provides in protecting the female against infanticide by other males.

In mountain gorillas (*Gorilla gorilla beringei*), protection from infanticide is also an important benefit females derive from associating with an adult male, and it may be the principal reason for female gregariousness in this species. Mountain gorilla groups generally consist of several females and their young and one breeding silverback male, although groups with several silverback males also occur (Stewart and Harcourt, 1987). Both males and females disperse from their natal groups, and social relationships among females, other than the mother-daughter bond, are not strongly developed. Mountain gorillas are folivores whose food is abundant and evenly dispersed, although there is spatial variability in its abundance and quality (Watts, 1984). Gorillas appear to "fall through the cracks" of Isbell's model in two respects: (1) within-group contest and scramble competition for food occurs, albeit at low levels despite the even dispersion of food; (2) there is a consistent, though subtle, linear female dominance hierarchy, with rank related to access of food priority and increased foraging efficiency (Watts, 1985; 1991). One reason for the lack of fit with Isbell's model may be that females also appear to compete for proximity to a silverback. While this aspect of female competition has not been directly assessed, some of the displacements over food, for example, may be a secondary consequence of jockeying for spatial proximity to a silverback (Watts, 1985). Indeed, the very low levels of interference competition suggest that group living imposes little nutritional cost on females, one easily outweighed by the benefits derived from male protection (Watts, 1985).

The importance of the silverback in protecting females from infanticide has been shown by Watts (1989). In eight of eleven well-documented cases, involving eight different males, infanticide resulted when the mother and infant were unaccompanied by a mature male, usually due to his death. This suggests that such vulnerable infants are certain to fall victim to infanticide. Infanticide shortens the interbirth interval, and a high probability exists that a female whose infant has been killed will mate with and transfer to the infanticidal male. In light of the slow reproductive rate of gorillas, these observations indicate that infanticide is on average a successful gorilla male reproductive strategy and that male protection is the primary advantage to gorilla females of group living (Watts, 1989).

Summary

In this brief review of the current state of primate socioecology, I have tried to show how a renewed focus on individual reproductive strategies has led to a clearer understanding of the causes of variation in social systems. At the same time, it should be clear that an understanding of only ultimate causes will be insufficient. We also need an understanding of the available behavioral options of a species, or breeding population, in pursuing evolved reproductive strategies under particular social and ecological conditions. Knowledge of the constraints imposed on behavior by the biology of a species, including life-history traits, such as brain and body size and anatomical and physiological specializations, will allow more accurate predictions about the influence of food resources, predators, or infanticidal males on primate populations. It will also enable a definition of the parameters of social plasticity. In this regard, experimentation with captive groups (Gore, 1991) will be particularly useful in that it permits the independent manipulation of socioecological variables thought to be important in structuring social relationships.

A holistic approach to behavior has been the central guiding philosophy, if not practice, of ethology. Both ethologists (Dawkins, 1989; Hinde, 1990; Stamps, 1991) and primatologists (Richard, 1985; Dunbar, 1988) examining social behavior have repeatedly called for the integrated study of species' biological and social dimensions. Yet, judging from the literature, most primatologists seem unprepared or reluctant to tackle the question of behavioral constraints. In the following section I suggest that, in keeping with the history of disciplinary fragmentation, this reluctance often stems from a misunderstanding of the nature and role of proximate mechanisms. As Hrdy has aptly stated, this fundamental misunderstanding of what a "biological basis for behavior actually means" has led to a situation where "we are all hapless passengers name-calling from the decks of ships as they pass in the night" (1990:25).

The Development of the Study
of Proximate Mechanisms

I want now to clarify what proximate mechanisms are and to discuss some historically based, common misconceptions about them that have haunted primatology and continue to do so. These misconceptions are about the reductionistic or deterministic nature of proximate mechanisms. As I will show, they are readily apparent in studies of the relationship between hormones and sexual behavior in nonhuman and human primates.

What Are Proximate Mechanisms?

The term *proximate mechanism* generally refers to the causes of individual behavior, particularly in connection with the physiological bases of behavior, such as neural or hormonal factors. However, the study of proximate mechanisms also commonly includes behavioral mechanisms, such as the effects of parental behavior on the characteristics of offspring, or the manner in which the behavior of an individual is affected by the behavior or morphology of conspecifics. Using primate social aggression as an example, we can consider two interrelated proximate mechanisms: (1) the relationship of endogenous hormones or neural activity to the performance of the behavior; and (2) the effects of social group size or individual dominance rank on the performance of aggression. A developmental perspective can be added to the investigation by examining the effects of maternal treatment, such as punitive versus permissive. Or in a twinning species, we could examine the uterine effects of a male fetus on a female fetus, due to increased exposure to testosterone. Though in this example the experimental focus varies, the proximate mechanisms, behavioral and physiological, are interrelated. Dominance rank, for example, may affect both hormone levels and expression of aggression, while, in turn, it may be affected by both in utero hormone exposure and type of mothering received. In this case, explication of the whole suite of proximate mechanisms gives us a more accurate picture of the raw materials available for behavioral evolution than we would get by focusing on any one aspect alone. In the remainder of this discussion I will treat proximate mechanisms in this broader sense.

History of the Study of Proximate Mechanisms

In primatology, the use of proximate mechanisms to explain social organization has been underappreciated for two reasons: they have been viewed as either inappropriately reductionistic or overly deterministic. Responsible, in part, is the historical separation of the social and biological sciences, and the resulting methodological and ideological tensions between the two disciplines. Such tensions become particularly apparent in discussions about the relationship between attributes of individuals, e.g., between their motivations or reproductive interests and the properties of social structure or organization (Hinde, 1983). In the disciplines most concerned with social phenomena—anthropology, psychology, and primatology—a central and continuing point of debate is whether all social phenomena can be explained fully by the actions of individuals without stipulating emergent phenomena at levels above the individual (Chadwick-Jones, 1987). In anthropology and related social sciences, Durkheim's dictum that social facts can be explained only by reference to other social facts continues to undermine efforts to establish links between the biological characteristics of individuals, the social system, and culture (Steklis and Walter, 1991). Such a view implies that lower levels of explanation (e.g., proximate mechanisms) have little or no explanatory power for higher-level phenomena like social structure. Any attempt, therefore, to translate between levels is considered inappropriately reductionistic. This view, however, stems from a fundamental misunderstanding of the role of proximate mechanisms in social behavior (Steklis and Walter, 1991).

Wilson and Lumsden (1991) have succinctly summarized this central problem in sociobiology as one of "holistic" versus "reductionistic" explanation. They state, "Holism implies emergence, the existence of processes at one level of organization that can be explained only with difficulty, if at all, by descriptions of processes occurring at the next level down" (1991:401-2). They suggest, therefore, that the problem is essentially one of method and knowledge:

> We need to expand our methodological focus to include all levels of organization and gather as much knowledge as possible at each level. The more we understand about processes at a lower level of organization (e.g., the individual organism), the less mysterious become higher levels of organization and holistic accounts. While it is tempting, for example, to describe the complex order and functioning of a fire ant colony as the result of "forces" acting at a higher level, distinct from the behavior of individuals, it can be explained as a "straightforward summation" of individual responses guided by relatively simple rules (Wilson and Lumsden, 1991).

Nevertheless, an understanding of how the food needs of the entire ant colony are met requires a shift in focus from the single worker to the level of interactions among workers. Holistic and reductionistic perspectives, Wilson and Lumsden argue, are not incongruent, but complementary. Taken in combination, they offer a more powerful perspective on social phenomena than either does alone (Hinde, 1990).

Proximate Mechanisms Today: Reductionism, Determinism, and the Problem of the Liberated Brain

This is a reasonable perspective, one with which most current primatologists would surely agree. But even "soft holists" (Wilson and Lumsden, 1991) might disagree about the deterministic qualities of proximate mechanisms, in particular, physiological mechanisms of behavior. Perhaps because of their perceived closer connection to genes, they are often and wrongly associated with having deterministic qualities. To suggest ways in which behavior might become released from physiological constraints and placed instead under the control of social and environmental stimuli is a strategy that appears to "reconcile" the flexibility of behavior with a deterministic view of physiological causation. In this view, cognitive evaluation of stimuli, especially in primates, plays the dominant causal role in behavior (Beach, 1947). Thinking of the relatively large-brained anthropoid primates (monkeys, apes, and humans) as relatively liberated from the constraints of physiology is consistent with this view. Moreover, primate behavior may be regarded as more liberated than the behavior of other mammals or non-mammalian vertebrates.

That this view is misguided is evident, for example, in considerations of the control mechanisms of primate reproductive behavior. For example, in his review of primate sexual behavior, Loy (1987) concludes from prosimians to humans that there is a progressive reduction of hormonal control over sexual behavior. He argues this reduction to be particularly significant in the transition from ape to human, with the development of a more obvious "situation-dependent" (Hrdy, 1981) female sexuality. This phylogenetic progressionist view is also apparent in recent evolutionary models of human sexuality that consider human behavior to be "soft wired" and female sexual receptivity "culturally mediated" (Szalay and Costello, 1991). Even prominent theorists of human sexuality who reject any juxtaposing of biology with culture, nevertheless fall prey to phylogenetic chauvinism. "[In] . . . subprimate species . . . sexuality and eroticism are governed under a hormonal dictatorship far more rigid than is the case in the primate species with their uniquely hypertrophied cerebral cortices" (Money, 1991:129). Similarly, Hrdy and Whitten, in their systematic review of primate sexual activity, say that "flexible receptivity" of primates is "quite distinctive" (1987:370), although they do not say why this trait should be more characteristic of primates than of other mammals.

Elsewhere, I have taken issue with this view of primate sexuality (Steklis and Whiteman, 1989) from a mechanistic perspective. Here,

I want to draw attention to the problems of this phylogenetic perspective. Many primate species do show situation-dependent, rather than hormone-dependent, receptivity; however, as pointed out by Aronson over thirty years ago (1959), the nature of the relationship between steroid hormones and sexual behavior is affected neither by the development of a large cerebral cortex (that is, a liberation effect) nor by phyletic position *per se*. The relaxation of physiological regulation is not in the purview of human, primate, or even mammalian species. Rather, study of the diversity of mechanisms controlling mating behavior shows us that the particular nature of physiological mechanisms for any vertebrate species depends on the history of selection pressures to meet specific ecological and social problems (Crews and Moore, 1986; Crews, 1987; Moore and Marler, 1988). The point the authors make is that it is incorrect to divide up animal species on the basis of whether or not their behavior is hormone dependent. Rather, they suggest, the expression of reproductive behavior must be coordinated with both internal and external (socioecological) cues. In all species, the effects of these stimuli on the behavior-regulating centers of the central nervous system can be mediated through a variety of pathways, of which sex steroids are but one.

Summary

In this section, I have discussed what proximate mechanisms are, described some of the historically based misconceptions about them, and given examples of their misapplication to the study of primate and human sexual behavior. In the following section, I will show that the modern study of the proximate causes of reproductive behavior, by taking into account their evolutionary diversity and plasticity, effectively undermine age-old criticisms of proximate explanations of behavior. Similarly, recent evolutionary perspectives on the mechanisms of intraspecific plasticity in social organization and behavioral dispositions point toward avenues of successful reunification of socioecology with the study of proximate mechanisms of behavior.

The Reunification of Proximate Mechanisms and Socioecology

Mechanisms of Reproduction and Socioecology

Crews and Moore (1986) demonstrate the linkage of behavioral reproductive strategies to reproductive mechanisms. Male white-crowned sparrows normally mate in the spring following gonadal maturation. While female mating behavior depends on sex steroid hormones, male mating behavior does not. Males are solely responsive to copulatory solicitations from females. At a different time of the reproductive cycle, males establish territories, and the requisite aggressive behavior is under tight hormonal control. Given the temporal separation of sexual and aggressive behaviors, their dependence on the same hormonal signal becomes dissociated. Since mating will only occur when the male is solicited by the female, this social cue provides sufficient temporal precision to the expression of mating. The initiation of territorial aggression, on the other hand, depends solely on the male and his gonadal hormones.

This example shows that the dependence of behavior on a hormonal signal or an environmental cue will vary with the relative utility of these signals for the appropriate temporal expression of behavior. It is no accident that in many species, gonadal steroids regulate mating and associated behaviors, such as intermale aggression, provided that fertility, mating, and related behaviors are temporally connected. In males of many bird species, for example, testosterone is most clearly linked to aggression associated with reproduction, territoriality, mate guarding, and dominance interactions, but not to anti-predator aggression (Wingfield et al., 1990). Notably, the authors show that blood testosterone levels are most closely tied to social challenges between males, or to periods of social instability such as during the formation of dominance relations.

Collectively, the above studies indicate that proximate mechanisms of behavior have been shaped by natural selection in very precise ways, especially where reproductive processes are concerned, in response to the social and ecological conditions that prevailed in the evolutionary past. This is elegantly demonstrated in the work of Gaulin and colleagues (Gaulin and Fitzgerald, 1989; Jacobs et al., 1990) on two closely related species of vole. They observed the relationship between their mating systems and sex differences in regard to spatial abilities. In the polygamous meadow vole, males range more widely than females and perform better on tests of spatial ability. In the monogamous pine vole, no such sex differences are apparent. Gaulin and colleagues found that, consistent with these behavioral and cognitive differences, only meadow vole males had larger hippocampi relative to overall brain size than females, a brain structure known, in many species, to play importantly in spatial learning. Also, among passerine birds, enlarged hippocampi correlate with specialized food-storing behaviors that rely on spatial memory (Krebs, 1990).

Proximate mechanisms appear to be sufficiently plastic, or

responsive, to evolutionary pressure, in that the multiplicity of reproductive strategies is matched by equal diversity in mechanisms that may exist, even between closely related species or among members of the same species (Crews, 1987). Given this evolutionary plasticity, we can expect frequent independent evolution of both similar and varied mechanisms that achieve the same functions (Crews, 1987). For example, reproductive suppression among females, a widespread strategy among mammals, is accomplished by diverse behavioral and physiological mechanisms (Wasser and Barash, 1983; McClintock, 1987). In dwarf mongoose, common marmoset, saddleback and cotton-top tamarins, subordinate female reproductive suppression is caused by the physiological effects on ovarian function of low status *per se*, not by behavioral effects of aggression received. Conversely, in golden lion tamarins and wolves, low status does not affect reproductive suppression, but aggressive interference does (Creel et al., 1992). Recall also, the gelada baboon females, whose reproductive suppression appears to be mediated through a combination of behavioral (aggression) and physiological effects (impaired ovarian function).

The perspectives presented here should lead to a new look at the relationship between proximate mechanisms and primate social behavior and point to fresh avenues for research. Primates present a wide array of behavioral reproductive strategies (Hrdy and Whitten, 1987). Not surprisingly, numerous species differences exist at the level of morphology and physiology (Klosterman et al., 1986). The job is to connect these two sets of data so that the combination makes sense socioecologically. To my knowledge, no one has attempted to consider, for example, the many different results of studies on the role of gonadal hormones and aggression in the light of species differences in reproductive strategies and mating systems. For example, both Wingfield et al. (1990) and Crews and Moore (1986) have provided new conceptual frameworks for the evaluation of species' differences in the biology of reproduction. These frameworks, based on general evolutionary principles, should hold for most vertebrate species and could, therefore, be productively tested in primates. Following Wingfield (1984), we could compare polygynous to monogamous primates in circulating testosterone levels and in the relative responsiveness of testosterone to social cues. Similarly, following Crews and Moore (1986), we might observe that gonadal steroids more directly regulate the aggressive and mating behaviors of seasonally breeding primates (squirrel monkeys, rhesus macaques) than they do those of primates that breed year round (vervet monkeys, stumptailed macaques).

Mechanisms of Social Plasticity
and Socioecology

I will now turn to the issue of behavioral plasticity, which has been implicit in this discussion all along. Although sometimes treatments of this topic become mired in the age-old nature versus nurture debate, most primatologists have moved beyond such simple dichotomies. Still, the issue is hardly resolved, and some old questions and biases persist. For example, to account for species differences in primate social organization most conceptions of the variables include species attributes or "species-specific constraints" (Richard 1985, see section III B). To think of such constraints in terms of body size and the many correlated life-history characteristics of a species is relatively easy (Harvey et al., 1987), although how these might constrain or determine behavior is poorly understood (Richard, 1985). More difficult has it been to understand how the behavior of individuals is constrained by evolved predispositions and how these, in turn, are related to social organization. Consequently, primatologists tend either to ignore the potential significance of evolved behavioral predispositions or to reduce them to genetics. Usually trotted out to illustrate the latter is the well-known (and worn) example of hamadryas-anubis baboon male differences in herding behavior (Richard, 1985).

Behavioral predispositions have little interested primatologists to explain variation in social organization, in part, because, like the phenomenon of genetic drift, this type of explanation can amount to a "waste basket" of failed explanations (Lott, 1991). In other words, if no other satisfactory (and more interesting) explanation can be found, then the phylogenetic history of a species, i.e., evolved behavioral predispositions, is invoked as the default explanation (Rodman, 1988). Thus, a species may be stuck with a particular behavior pattern that is immune to changes in socioecological conditions, and while it may or may not be adaptive in present circumstances, it also has not been removed by selection. Since closely related species occupying diverse habitats are more likely to reveal behaviors resulting from phylogenetic inertia, comparative socioecologists have tried to control for this possible "confound" in their attempt to determine what aspects of social organization reflect adaptations to ecology (Clutton-Brock and Harvey, 1977). Using similar logic, experimental studies that vary ecological factors have tried to identify phylogenetic factors in social organization; however, to isolate such historical effects from other possibilities has remained exceedingly difficult (Berger, 1988).

I suspect another reason why to ignore behavioral predispositions (or phylogenetic history) has been easy is that primatologists tend to think of most behavioral differences, either between individuals or populations, as resulting from the ability to make short-term, flexible behavioral adjustments to changing social or ecological conditions. This flexibility is achieved through the well-developed learning and cognitive abilities enabled by the evolution of large, complex brains. Learning and cognitive abilities in primates (and other social mammals) undoubtedly are related to solving socioenvironmental problems (Essock-Vitale and Seyfarth, 1987). Troublesome, however, is the conclusion that only large-brained vertebrates are capable of responding flexibly to changing environments or that, in order to be flexible, behaviors must be remote from the influence of genes. I will not go into detail here on the fallacies of this construction; they have been presented ably elsewhere (Tierney, 1986). Only will I add the observation that proximate mechanisms, including those resulting in behavioral predispositions or constraints, need not be genetically based to be heritable and thus to be of evolutionary significance (Stamps, 1991).

One way to grapple with the problem of behavioral dispositions and social plasticity is to change our perspective. We need to think of social plasticity itself as an evolved adaptation, rather than the consequence of large brains or reduced genetic input. Lott's (1991) broad review and synthesis of the causes of intraspecific variation in vertebrate social systems points to the importance of evolved behavioral dispositions as a key means of social plasticity. Further, the demonstration that social plasticity is common, but not universal, among vertebrates indicates the mechanisms of social change themselves to be evolutionary adaptations. Let us examine, in turn, Lott's reasoning behind each of these propositions.

If we can agree that social systems result from the behavioral interactions among individuals or individual "social strategies" under particular ecological and demographic circumstances, then it follows that a change in a social system results from alterations in the behavior of individuals. Therefore, we must look for mechanisms that alter the behavioral dispositions of individuals. Lott suggests that, because organisms must assess environmental, i.e., social and ecological, attributes so as to direct behavioral change, the proximate mechanisms of social plasticity at minimum must include assessment and behavior-regulating properties that, on average, lead to adaptive change. Assessment mechanisms can include learning and conditioning abilities and perceptual or attentional biases. We can expect these to be structured according to the particular set of socioecological problems that a species commonly encountered during its evolutionary history. Moreover, assessment need not involve complex cognitive operations. Indeed,

Lott argues that high intellectual function is not a common, nor particularly useful, mechanism for social system plasticity. Animals assess environmental features, such as mate or food-patch quality, by a variety of reliably associated, but indirect, cues— for example, body size or color brightness. In addition, animals may use direct cues to assess, for example, the competitive abilities of conspecifics (Stamps, 1991). This view of the mechanisms responsible for social plasticity accords with the widespread occurrence of intraspecific variation in social systems, irrespective of the degree of development of learning or cognitive abilities.

Social plasticity, however, is neither universal, nor always beneficial. This idea should not be surprising if we think of the mechanisms of social plasticity themselves as evolutionary adaptations. To predict the circumstances that might favor selection for or against social plasticity, costs and benefits have to be understood. On the benefit side, Lott proposes the ability to exploit new environmental opportunities that might result from environmental changes, such as different availability of resources or density of mates. Potential costs incurred include time spent in assessment and the risk of assessment error. Lott suggests that, in general, social plasticity should correlate positively with niche breadth: species that are omnivorous, have a low intrinsic rate of reproduction (K-selected), and inhabit unstable environments should be socially plastic. It will be interesting to see whether these predictions hold when primate species that show varying degrees of intraspecific variation in social organization are compared.

Beyond these general predictions about the incidence of social plasticity across species lie the limitations imposed on the social plasticity of a given species by interrelated social and biological characteristics. Such a consideration follows from the simple observation that a social "system" results from the interplay (and mutual constraints) among the behavioral and biological characteristics of individuals. We well know that body size affects energetic and, hence, behavioral constraints. For example, territoriality is uncommon in large-bodied herbivores because it is not energetically feasible (Lott, 1991). Body size may also constrain social relationships. Pereira (1992) has suggested intriguingly that the degree of sexual dimorphism in cercopithecines determines species differences in the pattern of rank acquisition. In rhesus and Japanese macaques and vervet monkeys, females promote dominance acquisition by all juvenile kin over group members of low-ranking matrilines. In baboons, however, females promote dominance acquisition only for female kin over female peers. The reasons are three: (1) adult male baboons are twice as large as females, and they invariably dominate all adult females; (2) female coalitions against male baboons are infrequent and ineffective; and (3) maternal status has no effect on male dominance relations, male migration patterns, or male reproductive success. All these preclude effective female agonistic tactics to influence male offspring dominance. Conversely, in macaques and vervets, where males are only 10 to 30 percent larger than females, some females can dominate some males either alone or by coalition and, thus, influence the dominant status of males, including sons. Consistent with this hypothesis, Pereira (1992) found that in the more dimorphic macaques, mothers help sons (but not daughters) maintain dominance over male peers born to low-ranking matrilines. In these species, males outrank females, but, unlike baboons, males dominate each other according to matrilineal rank.

Mechanisms of Behavioral Dispositions and Socioecology

Because the characteristics of individuals determine social systems, Lott (1991) has argued that selection acts on behavioral dispositions, not on social systems *per se*. This suggests that species-typical dispositions account for species differences in social organization. As for other phenotypic traits, a range of individual dispositions comprises each species, providing the basis for intraspecific variation in social organization. The possibility of species-typical dispositions, or temperaments, raises the important question of their relation to particular forms of social organization and ecological circumstances. In addition, the potential connection of dispositions to neural and physiological bases and also to social organization and ecology provides an exciting opportunity for linking proximate and ultimate causes of social organization.

We will now consider species-typical dispositions and their physiological bases. In our discussion of socioecology, we saw that the nature of intragroup relationships can be predictably related to ecological circumstances. That is, in those species where ecological conditions have favored cohesive grouping and strong within-group competition or dominance hierarchies, we can expect the appropriate behavioral dispositions and supporting physiological mechanisms to have evolved. De Waal's detailed comparative studies of dominance behaviors among primates (1989) address the nature of the behavioral dispositions that might be involved. According to de Waal, the flipside of competition and ritualized dominance relationships consists of mechanisms of tension regulation, e.g., social tolerance and conflict resolution. In fact, mechanisms for reconciliation have evolved in direct proportion to the value placed on the social relationship. That is to say, degrees of despotic or tyrannical behavior will be constrained because of the benefits that dominant animals derive from relationships with subordinates. Recall that among wild chimps, males form strong social bonds, while females do not. However, when competition for food is reduced, as at the Arnhem Zoo (the Netherlands) colony, females form coalitions and bond with each other, suggesting that, despite their relative solitariness in the wild, this aspect of their behavior is plastic. Interestingly, in this circumstance, we find no consistent female dominance hierarchy, no ritualized status displays, and no conflict resolution or reconciliation, as we do among male chimps. De Waal thus suggests that, while female chimpanzees have the potential for social bonding, they lack the mechanisms for coping with competition. Consistent with the pattern of female relationships in the wild, female chimps either lost or never evolved the "mechanisms of social dominance" comparable to those of males.

The concept of "species-typical temperaments" has been employed to distinguish the "dominance styles" of different macaque species (de Waal and Luttrell, 1989). While many species show strict linear or "formal" dominance hierarchies, they diverge in how they enforce status relationships in the face of competition. De Waal suggests that captive groups of congeneric species are ideal for revealing such differences in temperament, given that the potential influences of "ecology," body size, or dimorphism are controlled. In comparing two groups of rhesus macaques to one group of stumptailed macaques, de Waal found that, in the formal aspects of the hierarchy, the three did not differ, but that the dominance styles varied according to species. Compared to the rhesus monkeys, stumptails have a "looser, more relaxed" style. They showed more low-intensity aggression, but also more socially positive (approaching and grooming) and conciliatory behavior, and greater tolerance of subordinates. Also, more of the stumptails' aggression was reciprocal, and approaches were equally often initiated by and directed to animals of different ranks. In contrast, the rhesus monkeys were more violent and less conciliatory and most behavior was directed down the hierarchy. De Waal suggests that these different dominance styles reflect different "social values" in response to yet unknown ecological differences. Although study of aggressive patterns in three macaques species (Thierry, 1985) confirms some of these differences, such as the more violent aggression among the rhesus, the conclusion that these are true species differences will have to await study of additional groups.

In recent years, much effort has been spent on tying the

behavioral aspects of dominance systems to physiology. The acquisition and loss of dominance rank have been linked to biochemical, endocrine, and neurochemical alterations in a number of primate species, including humans. Most studies concern the interrelationships between steroid hormones and the aggressive interactions that accompany rank establishment or instability (Bernstein et al., 1983). Relatively fewer studies have addressed the causes and consequences of stable dominance rank. Unfortunately, virtually all these studies have examined male dominance relationships, making it difficult to tie the results to socioecological models of female-female competition. Nevertheless, the results of some of these studies interest us here because they point to a proximate basis of dominance relationships, independent of aggressive interactions, which may turn out to apply to female dominance relationships.

The physiological correlates of male dominance rank have been most clearly demonstrated in a series of elegant studies on captive groups of vervet monkeys (*Cercopithecus aethiops sabaeus*). (For a review, see McGuire et al., 1983; Raleigh et al., 1983, 1984.) This corpus of work shows that high male status correlates with high levels of whole blood serotonin (WBS), levels that rise and fall with dominance rank. Furthermore, dominant vervet monkeys differ from subordinate vervets both in social behavior profile and in their responses to a variety of test stimuli, all of which are related to brain serotonergic activity.

The relationship of WBS to dominance status appears to be influenced by species' differences in social structure. Among squirrel monkeys, high WBS levels correlate with high dominance rank (Steklis et al., 1986). In humans, they correlate with a "disposition to seek power": aggressiveness, competitiveness, and drive (Madsen, 1985). However, WBS and status are not related in *Macaca nemestrina* (Bowden et al., 1989) or in *M. fascicularis* (Shively et al., 1991). In the latter study, only those dominant males that initiated grooming frequently had higher WBS. Shively et al. speculate that in *M. fascicularis* the relatively greater importance of coalitions to dominance interactions (entailing grooming as indicators of reassurance and reconciliation) accounts for the more complex association between their WBS and status. Further, they suggest that the importance this species attaches to alliance formation, friendly relationships, and the more common appeasement gestures is consistent with the less-frequent occurrence of vervet multi-male groups. This suggestion recalls our earlier discussion about species' differences in valuing social relationships, despite similarity in formal dominance relationships (de Waal, 1989), and it adds a measure of subtlety to both social organization and physiology. But it will have to be further tested by examining WBS, status, and behavior interrelationships in species with similar formal dominance relations but differing qualities of male social relationships.

Species differences in behavioral dispositions, or temperaments, have also been linked to physiological differences. Behavioral and adrenocortical responses to stressful situations vary for captive rhesus, bonnet, and crabeating macaques (Clarke et al., 1988). In this study, for example, crabeaters, who showed the greatest signs of behavioral distress, also showed the highest corticosteroid levels following exposure to stress. This fact suggests that species differences exist in the mode of response to environmental stimuli. These three species normally live in multi-male social groups with dominance hierarchies and breed seasonally, but they vary in the nature of their intragroup interaction patterns (dominance styles, for example). The differences in physiological measures obtained are no doubt real, but we do not know which social stimuli would produce similar physiological responses.

What we need are studies of species-typical temperament and their physiological bases, studies that test predictions about ecologically relevant dimensions of social behavior. I offer the following questions and speculations solely to stimulate thinking along these lines. Are there differences in temperament between species with despotic versus egalitarian social relationships, and between female-bonded versus nonfemale-bonded societies? If in species with dominance hierarchies the cost of assessment and establishment of new social relationships results in "social inertia" (Stamps, 1991), then we might predict such species (frugivores) to be more xenophobic than ones without dominance hierarchies (folivores). Xenophobic responses, in turn, may be predictably related to dispositions, and physiological responses related to social stimuli (strange versus familiar). Are male versus female dispersion patterns associated with sex differences in dispositional or emotionality profiles and also with concomitant sex differences in physiological response to novel situations?

Much effort has been expended in examining how socioecological pressures have shaped the organization of the primate brain. Here, too, our new understanding of the selection pressures responsible for social evolution can play a major role in refining hypotheses. Recent comparative neuroanatomical studies of the primate brain conjoined with in-depth neurophysiological studies on particular species point to ways in which we can integrate ecological, evolutionary, physiological, and neuroanatomical information (Harvey and Krebs, 1990). Once again, this literature is too large to review here, but I want to point to some exciting trends in this line of research that combine the skills of neuroscientists and primatologists.

Our understanding of the neural basis of birdsong, food storage in birds, or vole spatial abilities indicates the evolutionary responsiveness of the vertebrate brain to social and ecological pressures. We might ask, therefore, if there are neural specializations that reflect the demands of primate social life. But to answer the question is difficult, because the neural machinery that evolved in the service of solving ecological problems may also serve in solving social problems and vice versa. The question has thus boiled down to whether ecological or social influences were primary in the evolution of both brain size and organization (Harvey and Krebs, 1990). One way to sort the two influences is to attempt to hold "ecology" constant, while examining the variation in a brain variable in relation to a measure of social complexity.

Two recent studies exemplify this approach. In the first study, Sawaguchi (1990) compared relative brain size (RBS) among forty-two congeneric groups of anthropoid primates in relation to diet (frugivorous, folivorous, and so on), stratification (terrestrial or arboreal), and social structure (solitary, single-male versus multi-male polygynous, and so on). When social structure and stratification were controlled, frugivores had larger RBSs than the folivores. Among frugivores, polygynous (multi-female) groups had larger RBSs than monogynous (single-female) groups, while terrestrial species in the same diet category had larger RBSs than the arboreal species. In a related study, Sawaguchi and Kudo (1990) examined similar ecological and social variables in relation to relative neocortical development. In both prosimians and frugivorous anthropoids, polygyny was associated with greater relative size of neocortex. The authors attribute these three differences to a complexity of social interactions that is greater in polygynous than that in monogynous groups. However, current socioecological models suggest levels of social complexity for multi-female groups that differ depending on the nature of female social relationships. Therefore, comparative brain studies might usefully divide polygynous groups according to the type of competitive female regime, in addition to the number of males.

Brothers has made a convincing case for the evolution of a specialized brain module of "social cognition," meaning "the processing of any information which culminates in the accurate perception of the dispositions and intentions of other individuals" (1990:28). She suggests that this ability to perceive "psychological facts" is unique to primates and best developed in species most closely related to ourselves: monkeys and apes. Primates, she suggests, infer the dispositions and intentions of others from such information as their movement, posture, facial expression and quality of vocalization, and from knowledge of their social and genealogical relationships. One of the features distinguishing social cognition from other kinds of cognition, Brothers argues, is the intimate tie between social information processing and affect, which has led to the evolution of a rich repertoire of emotional experience. Indeed, it is now well established that particular parts of the primate brain—the orbitofrontal cortex, temporal pole cortex, and amygdaloid nucleus—are specialized to process information about conspecifics and to connect such information with appropriate affective experience (Steklis and Kling, 1985). As reviewed by Brothers, the temporal lobe and amygdala accomplish these tasks in part by specialized processing of information on the face, including its expression of emotion and direction of gaze and facial identity, i.e., Who is it? and, What is this about?

Brothers's proposal for a neural module of social cognition fits well with our current conception of social group members as social strategists. As they pursue reproductive strategies in a social context, such strategies reduce to social strategies—making the best of circumstances. Thus, we may reasonably postulate the evolution of specialized brain mechanisms for drawing inferences about the capacities and qualities of group members, including their dispositions and intentions. Behavioral experiments (Dittrich, 1990) show that macaques discriminate among expressions of emotion, and that, in keeping with neurophysiological results, they process conspecific faces as "gestalt" images (recognition unaffected by orientation, color, size). While some of these processes need not necessarily be of a cognitive nature, behavioral evidence indicates that cognition, or "social intelligence," is involved (Kummer, 1978; Kummer, et al. 1990). This identifying of specific neuroanatomical structures, other than the cerebral cortex, involved in social cognition points to future research directions for comparative neuroanatomists.

In the final paragraphs of this section I want to explore an additional implication of the social cognition model, namely, that inferences about conspecifics' dispositions are reliable. I have already said that the existence of evolved, species-typical behavioral dispositions, like other phenotypic traits, implies a frequency distribution of dispositions in the population—a demography of behavioral dispositions. Again, as in the case of other phenotypic traits, if the trait is heritable and visible to selection, then the shape of the trait's distribution will be determined by the net impact of selection pressures. We must keep in mind that, as Stamps (1991) notes, the evolution of phenotypic traits can occur by routes other than genetic inheritance. What is required is both individual variation in the trait and a consistent relationship between the trait in parents and offspring independent of common environment.

What is the evidence for individual dispositions, temperament, or personality traits in primate groups, and how stable are these traits? First, these terms are simply short-hand descriptions for the overall attributes of an individual. Efforts have been made to objectify and quantify these impressions of individually distinctive attributes, either by rating subjects along dimensions of emotionality-personality, e.g., aggressive, controlled, depressed (all derived from human emotions-personality theory [Buirski et al., 1978; Martau et al., 1985]) or by using a rating scale comprised of adjectives that describe human personality, e.g., excitable, eccentric, insecure (Stevenson-Hinde and Zunz, 1978; Stevenson-Hinde et al., 1980). General agreement exists that individual differences in temperament, evident early in development, are strongly influenced by biology (Schneider et al., 1991). Indeed, in nonhuman primates, such differences as fearfulness or anxiety are related to neurochemical and hormonal profiles. Even under conditions of cross-fostering, profiles remain more like those of their biological parents than of their adoptive mothers (Higley and Suomi, 1989). Similarly, different styles of dominance and subordinance among wild adult male baboons are associated with distinct endocrine responses to social stress (Sapolsky, 1991).

This glimpse at primate personality indicates that temperaments and dispositions are stable, biologically based attributes of individuals. These attributes, therefore, are not just constructs or abstractions, but are relied on by nonhuman primates to assess the value of social partners. Whether such individual differences are the byproduct of selection for species-typical temperaments, or whether they themselves are the consequence of selection remains speculative. In reviewing the evolution of human personality, Buss (1991) suggests that the consistency across time, contexts, and cultures of a basic set of human personality dimensions indicates that these dimensions result from a shared set of evolved psychological mechanisms, that is, from "human nature." Like our view of the species-typical dispositions of nonhuman primates, we can see human psychological mechanisms as the solutions to common social adaptive problems that ancestral humans confronted. Buss proposes that personality differences, such as "hierarchical proclivities," "willingness to cooperate," "ability to handle stress," and so on, represent evolved, alternative behavioral strategies. He suggests that, in light of such individual differences, selection would have favored "those whose capacity to discern the differences enabled them to increase their inclusive fitness" (1991:473). If we assume that similar processes are responsible for the evolution of personality differences in nonhuman primates, we can expect the degree of individual distinctiveness to vary with measures of social differentiation, such as hierarchies and social alliances.

I realize that I have covered considerable ground in this section, conceptually and taxonomically, so as to highlight misconceptions about the nature of proximate mechanisms and social plasticity. The theoretical principles of socioecology apply to a wide range of social vertebrates. Indeed, most were and continue to be derived from studies of nonprimate species (especially birds). Similarly, nonprimate species have been the source of data and theory concerning the evolution of proximate mechanisms. As in socioecology, the principles imported into primatology have helped clarify the nature of proximate causes and will continue to have heuristic value.

Socioecology from the Bottom Up
A Schematic Summary

I have traced the largely separate developments of primate socioecology and the study of proximate mechanisms of behavior. I have made evident that socioecological models, through their focus on the adaptive behavioral strategies of individuals, have directed attention toward the proximate causes of behavior. Study of the mechanisms underlying behavior have shown how these have been shaped by natural selection in order to enable organisms to respond adaptively to social and ecological stimuli. In order to construct a model of socioecology "from the bottom up" the task still remains of connecting the biological characteristics of individuals to observed social structures.

The figure below illustrates, in a simplified manner, how primate social structure might be derived from the bottom up. It is an attempt to schematically summarize the causal factors of a given population's social structure at any one time. The scheme accounts for both between- and within-species variation in social structure. The figure is organized so as to draw the reader's attention to a series of factors (from phylogeny to ecology) that successively constrain the finally observed range of social structures. One might think of this scheme as a Tupperware model of the causes of social structure, with phylogeny as the largest bowl containing or constraining all the others, among which the smallest represents an eventually observed social structure. Let us examine the causes of social structure from phylogeny upward.

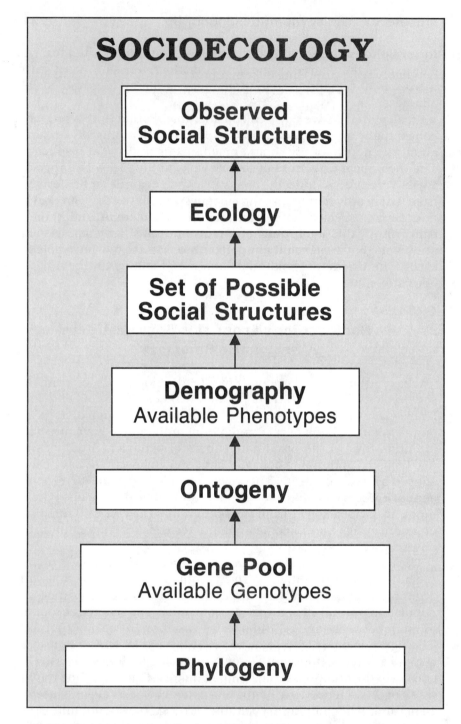

SOCIOECOLOGY

Observed Social Structures

↑

Ecology

↑

Set of Possible Social Structures

↑

Demography
Available Phenotypes

↑

Ontogeny

↑

Gene Pool
Available Genotypes

↑

Phylogeny

This figure schematically illustrates the causal factors of a given population's social structure at any one point in time. The scheme accounts for both between- and within-species variation in social structure.

Phylogeny

The genotypes comprising a population at any time partly result from the population's evolutionary history. These provide, therefore, very basic constraints on the range of possible phenotypes. As I have said, phylogeny (or phylogenetic inertia) is invoked when other factors do not appear to account for variation in phenotype.

Gene Pool

The actual genotypes comprising a population, that is, its gene pool, are the result of current sexual recombination (matings) of the historically available genetic material and any new variations (mutations) that are present after recombination. Any set of zygotes formed from this genetic material thus finally determines the range of possible phenotypes for any one generation.

Ontogeny

This factor represents the full range of developmental influences on a genotype. However, as the responsiveness of the genotype to environmental stimuli is itself determined by genetic properties (e.g., genes that can be turned "on" and "off"), genotype constrains ontogenetic influence. Such constraints are evident, for example, in species-typical (ranges of) growth and behavioral developmental trajectories. Although organisms respond selectively to the myriad of possible environmental stimuli (i.e., they define environmental stimuli), the degree of variation in environmental stimuli will relate to phenotypic variability.

Demography

I want to extend the usual definition of demography (age/sex structure of a population) to include behavioral phenotypes. I have already suggested that many examples of demography act as an independent cause of social variation, a suggestion that follows from our considering social organization as an outcome of patterns of social interaction. However, these examples have necessarily focused on the following effects on the social system: varying availability for social interaction of age mates; potential sex partners; reproductive rivals or other competitors. Now I suggest that, given that personality traits are stable and heritable, their distribution in the population (independent of age and sex) may account for some portion of the variance in social organization whenever other demographic variables are held constant. Thus, the demographic structure of a population that includes the set of behavioral dispositions should be the one that further defines the range of possible social structures.

Ecology

Local ecological conditions (predation pressure, food availability, density of competitors) will influence which of the potential social structures—given all the previous constraints—is expressed. Because we begin with phylogeny, we can expect some degree of fit between phenotype, social organization, and ecology, although this may not be apparent in all current situations, given that this fit was forged in the past. Furthermore, the "goodness of fit" will vary with species: individuals of K-selected species that evolved in unstable environments also have so evolved the capacity to vary their behavioral strategies that populations of the same species may pursue different (social) solutions to similar ecological circumstances.

Concluding Thoughts

My main goal throughout this chapter has been to show the interrelationship between the proximate and ultimate causes of primate behavior. It is only through their combined consideration that we can fully appreciate the evolutionary processes responsible for the diversity in primate social organization, including that of humans. As the research of both behaviorists and biologists is converging on the same questions about the origin and functions of animal social behavior, the time has come to bring traditional ethology out of the closet and into the mainstream of primatology. Then, perhaps, the "promise of primatology" will be fulfilled.

Rivalry, Resolution, and the Individual
Cooperation among Male Langur Monkeys

Phyllis Dolhinow and Mark A. Taff

Phyllis Dolhinow (University of California, Berkeley) is one of the founders of modern primate field research and has written several volumes in the field. Her book, Primates, Studies in Adaptation and Variability *(1968), was one of the first to look at the relationship between behavior and ecological variables. Here she is joined by Mark Taff (University of California, Berkeley) to discuss conflict resolution in langur monkeys.*

Perspectives and information gained from studies of living nonhuman primates have helped us identify and reconstruct important milestones in human evolution (Dahlberg, 1981; Fedigan, 1986; Harding and Teleki, 1981; Isaac, 1978; Tanner, 1981; Washburn, 1973a, 1973b; Washburn and Hamburg, 1965). Our understanding of the dynamics of nonhuman primate social relations, or networks, and of their interconnectedness with the animal's environment continues to grow, and thus continues to inform us about the most likely paths of human evolution. At the same time, the more we learn about hominoid behavior, the more we appreciate its richness and variety. Yet, despite the gains in insight provided by studies of primates, Washburn could not have been more correct when he wrote in the early 1950s that "there is nothing we do today which will not be done better tomorrow" (1951:304).

As we acquire knowledge we constantly refocus or recast our questions and proceed to construct "best possible" answers, knowing that what is probable now may soon be re- or deconstructed as the field of primatology matures. For example, early studies of baboons in East Africa constructed a scenario of rigidly male-dominated troops within which the tempo and format of daily life was based mainly on male relations (DeVore, 1965; DeVore and Washburn, 1963; Washburn and DeVore, 1961). Because dominance hierarchies were known to structure groups, we thought that dominance was the means by which issues of rivalry were resolved and that aggression was the emotion of choice to express social control. This androcentric portrayal of primate behavior, baboon behavior, expressed the social science of the 1950s and early 1960s. It also reflected some of the constraints stemming from relatively short periods of observation.

More than thirty years later, we join with others (for example, see Bateson, 1984; Hinde, 1987) in taking a close look into what continues to be an important aspect of primate behavior, the dynamics of rivalry; and, along with them, our focus turns to cooperation and supportive interactive behaviors as critical elements of conflict resolution. Our examples are drawn from a long-term study of a colony of Indian langur monkeys (*Presbytis entellus*). We report specifically on male-male relations, which constitute complex social networks expressed in behaviors ranging from strongly affiliative to extremely hostile. In considering male rivalry and the resolution of hostility, we draw attention to the tremendous effect single individuals can have on the outcomes of conflict, whether that conflict involves two animals or all members of the group.

Primatologists have recognized for many decades that great variability exists among the individuals of every primate species and that we must consider the behavior of individuals in order to understand group dynamics (Dolhinow and Taff, 1990; Goodall, 1986; Fedigan, 1982). However, we must resist the convenient tendency to de-emphasize the need to follow individuals for their entire lifetimes, or better yet, for generations. If we document the range and nature of individual variability for only a few years, as happens in most studies, we will probably not only chance missing critical events, but we will also not witness processes, such as those during development (epigenesis), that result in the patterns of behavior needed for survival. To understand behavior from an evolutionary perspective, we must understand what it is that will affect the individual's survivorship and reproductive potential — in other words, the individual's fitness. However, since fitness is not a property of an individual, but rather the interaction between a specific phenotype and the environment, as the environment changes, so does fitness. To appreciate how the behavior of our primate relatives, both living and extinct, has been shaped over time requires an understanding of the day to day events that additively affect the individual's fitness.

Behavior is generated by individuals in a context that includes the biotic (self, conspecifics, and others) and abiotic (the environment). In addition, every animal has a history that affects current behavior (for example, see Oyama, 1985), and careful consideration has been given to the interaction of experience and perception on individual behavior (Cheney and Seyfarth, 1990). If our study of behavior is not epigenetic, that is, throughout the lifetime, we run the risk of ignoring the tremendous significance of experiences during development (Jamieson, 1986). Even with our knowledge of much that has happened from birth onward in the lives of our subjects, we are far from being able to predict what decisions animals will make and how these decisions will affect the nature of the social group.

We hold much in common with our close relations, the living nonhuman primates, including sensory perceptions, emotions, a propensity for life in complex social contexts, and many of the same challenges of staying alive and remaining safe from predators and the elements. These similarities allow us to appreciate much that happens in the nonhuman primate's life. However, the major difference between human and nonhuman primates — and it is an overwhelmingly important one — is cognitive behavior, including culture based on human language, a sound code, made possible by evolutionary recent changes in our brains (Washburn, 1982). Despite all the insight we gain because of phylogenetic similarities and our investigations into the lives and perceptions of the primates, we still do not know what information and experiences inform an animal's actions. Often we can only guess at an animal's "motivation" for its actions. When we do guess, and at times doing so can be both appropriate and productive, it should always be with the caveat that we may be wrong. Animals cannot tell us why they behave as they do, and in our efforts to infer their reasoning from what they do or do not do in many instances, we must be extremely cautious not to commit the error of affirming the consequent. It is all too easy to move from presumed effect to ostensible cause, offering the very actions we seek to understand as evidence for what we consider to be the cause of the behavior.

With all the above cautions in mind we look in our case study at the behavior and changing social relations over a three-year period within a captive all-male group of langur monkeys. Stable social groups composed only of males are typical for this species, and males from as early as two years usually spend portions of their life in an all-male group. Relatively little is known about the social dynamics within these male groups from field observation (Mohnot,

1984; Moore, 1985; Rajpurohit, 1991), but we have collected important data from our captive colony.

Our observations suggest two major patterns of organization: one is relatively stable for long periods; the other is based on short periods of changing social relations with rapidly shifting patterns of power and alliances. After consideration of some variables that contribute to change in the group, we conclude that male group structure changes according to both internal and external stimuli. Ultimately, individuals determine the direction and extent of social change as they interact; however, the duration of stasis or the magnitude of change is based much less on conflict than on the nature and strength of cooperation among pairs of animals in the

Two adult male langur monkeys sit side by side.

group. Because of alternating periods of stability and change, the picture of group social relations will vary dramatically depending on what point in the life history of the male group is studied.

Langur Males in India

Patterns of langur monkey group structure and social organization are exceedingly varied throughout India (see Taff, 1990 for a review). Social groups may contain a single or multiple adult male(s) with a number of adult females and immatures. Associated with these mixed-sex groups are generally smaller groups composed only of males. Male groups have been reported for all major field study locations in India except for sites in the Himalayas. Descriptions of social relations in these groups have been extremely variable from one location to another and also for individual locations. While we acknowledge the value of understanding male group formation and dynamics, our records of the behavior of male groups are meager. Some observations taken from field reports are discussed below.

Male groups in some locations in India are described as loose aggregations with little evidence of a dominance structure (Sugiyama, 1967), while other groups are reported to have stable linear hierarchies (Hrdy, 1977; Moore, 1985; Newton, 1987; Sommer, 1988). In a few instances both patterns occur at a single location (Mohnot, 1984). To date the most complete description of male group behavior is provided by J. Moore (1985). His study, conducted at Ranthambore, India, suggests that the organization of males into smaller subgroups was an important feature of all male groups. He based his identification of subgroups on their members' tendency to sit near one another, groom, have few displacements, and exhibit a general cooperativeness. Although he had to estimate ages because his study was a short one, he concluded that the subgroups were based roughly on age—that younger adult males tended to join one unit and older males went into the other. Although the genealogical relations of males in the male groups were unknown, Moore suggested that kinship was a very important organizational variable in male group dynamics. This may be one of those reasonable expectations that will not be supported by the data when more detailed information is gained. The answers will be reliable only after the male groups, *and* the two-sex groups from which the males come, are observed for long enough to know kin relations, group of origin, and dates of transfer.

Data from the field, then, suggest that male groups are the major location of life for most, if not all, males at some time in their lives and that these groups play an important role in the social ecology of the species. During the course of adulthood, males probably move in and out of bisexual groups, interacting with both familiar and unfamiliar adults according to strategies of both cooperation and conflict originating in the natal group and practiced among males in male groups. To fill out the picture of life in a male group we turn now to our detailed long-term study of male relations in a colony.

The Colony Male Group

This chapter is based on a portion of our study of a langur monkey colony established in 1972. Four generations have been monitored since the formation of the colony, and from this extensive body of data we discuss a three-year period of observations of an all-male group. In addition to 307 hours of focal animal samples (Altmann, 1974), we draw on 380 hours of ad lib observations (see Taff, 1990 for details). We selected these data from the twenty-year record because this block of observations witnesses social changes often exceedingly rapid, unexpected, and, in retrospect, opportunistic. We take a unique look at the details of male-male relations and gain insights into potential constraints affecting male rivalry and alliance in both all-male and two-sex groups.

During the three-year segment under consideration, two major patterns of organization were recorded for the male group. The first pattern of organization involved was a relatively stable network of relations in which each member of the group occupied a clearly defined place. Males in the male group typically aligned themselves in pairs or dyads, which have been described by Taff (1990). These relations also have been described as coalitions which are special dyadic relations that appeared to be mutually beneficial. Although the term "coalition" is sometimes narrowly applied to mutual assistance during agonistic encounters, it is applicable in a broader sense to the dyadic interactions seen in the male group. Three aspects of the pattern of dyadic relations suggest the dyads are appropriately described as coalitions: (1) these relatively stable pair relations were not random and were actively maintained by at least one of the partners; (2) these relations appeared to benefit one or both of the males; (3) though rare, there were examples of a male's receiving aid from his partner during an aggressive encounter. The pairs or dyads are also easily identified by their spacing patterns, mutual grooming, and visual monitoring of each other.

The second pattern of organization for the male group is identified as an interval of instability of social relations with shifting patterns in both conflict resolution and access to objects, partners, and

Two adult female langur monkeys groom as they relax in the afternoon sun.

places. The social structure of the male group changed as a result of events originating from within and/or outside the group. Two events we describe that contributed to short intervals of instability and change included: (1) the introduction of a new member into the male group; and (2) activity in the nearby two-sex social groups. Perhaps the most outstanding characteristic of the langur males during the period of rapid social upheaval was a very pervasive opportunism, demonstrated by every male in the group at one time or another. This opportunism often occurred in the absence of any discernable stimulus or action by any other animal. It took the form of actions, including ones new to that individual, and it also was observed in the seemingly instantaneous formation of alliances with individuals formerly treated as rivals.

In September of 1985, six males, aged five to nine years, were put together to form a social unit. The males came from three different social groups and there were varying degrees of familiarity among them. The group was placed in an outdoor runway (23 m x 6 m x 4.6 m) with an attached, heated sleeping room. All males were introduced on the same day and immediately there was competition among them, with the formation of temporary associations. There was, however, no clear indication of individual ranking. All males fought, though four were responsible for most of the aggression. On the first day these four formed two loose coalitions and the pairs competed against one another. On the third day fighting had subsided and the males established patterns that would characterize them during the next fourteen months. It took five days for the group to stabilize during which time three pairs were established. The pairs were males 1(M1) and 2(M2), 3(M3) and 4(M4), 5(M5) and 6(M6), aged nine, six, seven, six, six, and five years respectively. There followed a two-and-a-half-year period of stability during which the composition of each pair remained unchanged.

The Big Change

In March of 1988, a five-year-old male (M7) was added to the group and, once again, it took five days for social relations to stabilize. However, with the addition of M7 there was then an odd number of animals, and this affected the pattern of male associations. After M7's introduction, the close bond between M1 and M2 weakened. Whereas the two dyads of Ms 3 and 4, and Ms 5 and 6 were unaffected, M1 remained alone much of the time. M7 spent a great deal of time with M2, although this was not a typical dyad in either frequency of interaction or support during agonistic interactions. M7 was definitely the lowest in rank of the entire group as measured by pairwise interactions in agonistic or competitive situations. The structure of the group was, then, M1, Ms 2 and 7, Ms 3 and 4, Ms 5 and 6. M1 maintained social control over each of the group males. He could displace all others, take favored food, and move or sit anywhere regardless of who was already there.

Five months after M7 joined the male group a series of events culminated in a dramatic change in group social relations. The initial stimulus was to come from commotion in the mixed-sex social groups housed near the male group. For management purposes, animals were being moved between the two-sex groups starting at approximately 0900. Since auditory and limited visual contact was possible, it may have had a disruptive effect on the male group. At 1445, the male group began loud vocalizations and very high energy displays, careening off walls and all structures in the enclosure. Ms 2, 3, and 4 attacked M1 and quickly cornered M1 in the sleeping room. The three aggressive animals ground their teeth loudly and M1 alarm barked continuously. The alarm bark is one of the most serious vocalizations a langur makes. It is an excellent indication that the animal is very alarmed or frightened. Alarm barking that lasts more than a few minutes is exceedingly rare, yet M1, who had never been recorded to emit an alarm bark, did so for more than thirty minutes. Not since the formation of the male group three years previously had M1 been in a position other than number one. Now he was in a situation where he was trapped by the physical presence of the three males acting in concert against him. This attack reflected the change in relations that had already occurred. The structure of the group had altered rapidly and radically.

The rate of aggression (including face threats, hair pulling, hitting, biting, chasing, and fighting) on the first day of the change was 37.6 bouts per hour. This rate compared to an average of 6.56 bouts per hour during the previous fourteen months of observation. The result of Day 1 fighting was a complete alteration of the relations of M1 with the other males. Three changes in the pattern of interactions between the males reflected this alteration. First, he was attacked by each of the others at least twice for a total of sixteen observed attacks in three hours. This was totally unexpected, based on observations from the preceding fourteen months. In more than three hundred hours of focal animal samples M1 had been threatened only twice, and these were noncontact threats. Second, M1 exhibited a sudden noninvolvement in the agonistic activities of the other males. It had been common for him to intervene in any rivalry expressed by other members of the group, but by the end of the day, M1 was exceedingly wary and often ran away to avoid confrontation. And third, M1's carriage and manner were markedly changed. He watched the others constantly with jerky, apparently nervous motions. His shoulders were hunched and his stride short and tense. Interestingly, all the previous characteristics of M1, his interventiveness, his confident, leisurely manner and relaxed posture were assumed by M2.

On Day 2 aggression dropped to 18.9 bouts per hour. "Opportunistic" groupings were formed for temporary advantage during agonistic exchanges between the males. In one instance M1 seized the chance to work in concert with Ms 5 and 6 against M4. M1 and M4 had previously shown very little affiliation; from day one of the change M1 was very aggressive to M4. However, this changed on Day 5. M1 acted with M3 jointly against M5, and M1 and M3 continued to cooperate. Then, because M3 and M4 were coalition partners, the result was a triad made of Ms 1, 3, and 4.

By Day 5 levels of aggression returned to nearly those of pre-change days and new patterns of affiliation were established. There was also a complete change in the relation of M1 with the other males. M2 was able to gain favored food items in the presence of any of the other males. M1 simply did not compete. The hostility that apparently existed between M1 and M4 seemed to change on the fifth day. Though conflict between these two males was evident at the end of Day 4, on Day 5 M1 and M4 were staying close to each other as they ate. Approximately ten minutes after the feeding ended, the first example of their cooperation, a joint displacement of M5, was recorded. While in fact this may not have been the first time the two males cooperated, it did appear to signify a turning point in their relation. After that time they continued to cooperate and sat together frequently. It was at this time that the fissioning of the group into two subunits occurred.

On Day 7, after much challenging and testing and many different combinations of alliance, a balance was reached. The group had effectively split into two units, each of which associated primarily with its own members. The M1 unit included Ms 1, 3, and 4 and the other, the M2 unit, included Ms 2, 5, 6, and 7. Members of each unit were usually found near each other and in different parts of the enclosure from the members of the other unit. Affiliating interactions were almost exclusively confined within the unit. The genesis of this new structuring, which amounted to a group split, although certainly based in part on earlier patterns of affiliation, can be seen in patterns of agonistic interactions that emerged during the intense aggression observed at the start of the change. Eventually the characteristics of each subgroup or unit differed. The M1 unit was less close-knit but relations were quite calm. In the other unit, M2's, M6 and M7 were rather aggressive toward each other. What apparently kept this latter unit together was a series of dyadic relations that ultimately intertwined. The central focus was M2, and the patterns of affiliations with M2 brought the four males into a network of relations that prompted its designation as a subgroup or unit.

Why the Change?

There were a number of ostensible causes. The apparently triggering stimulus was the noise and commotion in the nearby two-sex groups. During the ensuing aggression the members of the male group determined that the previously unchallenged M1 was no longer in control. He was found to be defeatable and, indeed, the others ganged up on him and defeated him. Subtle indications had been noted earlier indicating that tension was building between M1 and M2. This was manifested by patterns of visual monitoring and spacing. When M7 was added to the group four months before the big change, it created a situation in which the members of the male group retested all relations and every individual doubtless reevaluated the abilities and other attributes of the other males. M7s appearance was followed by immediate changes in the relations between the established dyads. Agonism increased between M6 and M7 which led to M7 interacting more with M2. There was then a decrease in affiliation between M1 and M2 with M1 spending more time with M3. As noted, there is more conflict among the males when there is an uneven number of animals in the group. Again, these observations illustrate how a series of smaller, often quite inconspicuous, events can result in an episode of major change. Understanding the change requires the identification and evaluation of as many of the smaller initiating stimuli as possible.

So, What Can Be Said about Rivalry and Its Resolution?

How then, are we to characterize langur male conflict and cooperation? Obviously, there are many ways an individual can respond in every situation, and the choices each individual makes depend on more factors than are known, or knowable, to the human observer. Each male acts according to his experiences and his evaluation of the situation at the moment (Curtin, 1981). Males may act or not, play it safe or seem to take chances, or remain apparently uninvolved. We do not presume to know what an animal wants, whether to win, to control, to relax, to stay out of the way, to be left alone, or something else from a long list of other outcomes. Keep in mind that an individual may have completely different goals in mind at different times. As to motivation, there are legions of questions and sadly few answers.

No animal can remain completely uninvolved when group relations change. Although some individuals may appear to be minimally affected by what goes on around them even when others are being loudly and boisterously aggressive, this nonparticipatory stance does not last long. Sooner or later, and to a greater or lesser degree, every member of the group will be pulled into the activity. It is then that an ability to recognize and to select among a number of possible responses at each choice point becomes critical. The fact

that there are probably a number of possible responses in every
situation means that every individual has an opportunity to
discover and then to monitor change in both his physical and social
environment. Mistakes will be made and there must be margin for
error. An important benefit from "error" lies in discovering change
and new solutions to problems. As Bernstein so aptly put it:

> You've got to be stupid in order to be smart . . . The genetic
> contribution to primate behavior may include a predisposition
> to vary responses to the same stimuli. Such variability produces
> a mix of "erroneous" responses to stimuli, but such errors may
> serve to monitor change in contingencies in the environment
> (1984:297).

After typically long periods of stability or stasis, male langur
group structure and social relations can change rapidly and
drastically. Certainly one important aspect of male behavior during
challenges to established relations is the ability to be opportunistic.
Our information suggests that previously established lines of
cooperation may diminish greatly during times of rapid change.
Established dyadic relations may be short-lived and clearly some,
if not most, are constructed opportunistically. This is in contrast
to periods of stability when some degree of cooperation is the rule.
In fact, the balance in social relations is maintained as a result of
cooperation. During the "big change" the males worked together
in actions that were, if not planned, at least coordinated and, at least
for the moment, were to the advantage of the participants. What
is interesting about the brief associations that formed during the
change is how opportunistic they were and how older patterns of
affiliative relations were quickly abandoned. Though past
experiences of interaction are important, what seems to be most
significant is the context of the action at the moment. It appears
that even long-standing patterns of alliance can disappear quickly
if an individual perceives the situation as calling for a reversal of
affiliation.

The behavior of individuals can influence the entire character of
a group. In the case presented here, the individual actions of M5
contributed greatly to the overall character of the group. His high
levels of aggression set the stage for the change and helped shape
what was to happen. The longitudinal nature of our research
provides us with knowledge of some past events in this individual's
life which may have contributed to his behavior. He had a number
of rather unusual factors in his history, including growing up in
a social group where he had no age cohort. Consequently, his many
hours of rough and tumble play were with adult playmates, and
he was subjected to more energetic wrestling than are most

immatures. He learned at an early age to manage adult rough play and endure aggression — he was forced to respond strongly. Obviously the situation is complex, but it is clear that one individual can produce a pattern of behaviors with a net valence that directly affects the character of a group.

At the very least, whether langur male group structure remains stable or changes is context dependent. Our data suggest a resolution to the apparent discrepancy concerning dominance hierarchies in male groups as described in reports from the field. In groups characterized by stable relations, foci of competition seem to be greatly reduced. As indicated above, social relations can remain unchanged for relatively long periods. On the other hand, social relations may become unstable and change can occur rapidly, be very intense, and execute powerful redefinitions of rank, cooperation, and association patterns. Different interpretations of what is happening in male groups may be functions of when, in the natural cycle of change, researchers observe these social units.

The Potential Costs of Rivalry

The incidents described in this chapter illustrate the subtlety and complexity of male relations. The reader may notice that in our descriptions of male relations we avoided describing them in terms of dominance. We did so because dominance rank or position is not an attribute of the individual: it is not a rank assigned as the result of the application of mathematics to competitive outcomes (Bernstein, 1981; Hinde, 1978). Motor strength and agility, competitive inclination, problem-solving skills, and social savvy are among the many intervening variables that influence competitive and affiliative outcomes; further, they certainly constitute important knowledge that each member of the group uses in all interactions including those in rivalry situations. An asymmetry of outcome — for example, if one individual succeeds in taking the prized object or asserts his ability to make another move from a position — is not proof of the existence of specific relations. If jumping to conclusions, we judge it so, we are most likely committing the error of affirming the consequent using the ostensible effect as proof of our presumed cause — the rank relationship. Importantly, we cannot predict what will happen in a situation, even knowing a great deal about all the actors. When we try to anticipate the response of an individual to certain factors in the physical and social environment, we must be cautious to formulate our predictions as probability statements. We cannot know the variables the individual uses to make its choices of action, or inaction, and must keep in mind that an animal may have a number of possible options.

Although cooperation serves to reduce the drastic effects of rivalry aggression — not only by diffusing the target of hostility among a number of animals, but also by reducing tension, as when one individual reassures another — the results of rivalry are apparent in captivity and they can range from threats with no physical contact to fighting that may result in serious wounds. The same situation pertains in the field where male rivalry and conflict generally occur with minimal damage to the participants. However, in some areas of India fatal consequences result from male rivalry. As mentioned above, male group transfer from an early age is a characteristic of the species. In populations where multi-male groups are common, as at Rajaji, males enter and leave two-sex groups during the mating season but with no significant fighting and certainly no loss of life (Laws and Vonder Haar Laws, 1984). In other areas such as Jodhpur, Mount Abu, Dharwar, Kanha, and possibly Harihar (Hrdy, 1974, 1977; Sommer, 1988; Sugiyama, 1965, 1967), the entrance of males from all-male groups into the two-sex groups can be accompanied by massive fighting and, at times, the death of group members (Boggess, 1979, 1980). Male rivalry as we observe it in these areas is a critical component in the aggression central to scenarios of rapid and aggressive "male takeovers"—the term applied to a male or males entering a two-sex group and assuming control of the resident group females. Some observers have asserted that the purpose of the invasion is to kill offspring of other males, labeled as "infanticide," inseminate the females, and become the "father" of all future infants (Hrdy, 1974; Agoramoorthy et al., 1988). Although the witnessing of males actually killing infants is rare — one estimate is nine of approximately fifty "killings" (Agoramoorthy and Mohnot, 1988) — the context of the deaths that have been recorded leaves many unanswered questions (Curtin and Dolhinow, 1978). These events are often described as a behavioral complex that ultimately leads to an increase in the male's reproductive fitness, thereby constituting an evolved reproductive strategy. The evidence needed to support this so-called strategy is lacking — we lack such critical data as paternity, pregnancy outcomes, and the immediate context of the "killing" including, remarkably enough, which animal (male or female) did the killing! In our search for the problems to which observed or assumed behaviors are the solution, we have made unwarranted generalizations. By a completely unwarranted leap from effect to cause, some consider fatal events such as "infanticide" to constitute a reproductive strategy for male success.

Unfortunately, langurs have been labeled as infanticidal as

though the behavior of a few individuals in some groups represented a norm for the species. The major locations at which infanticide is alleged to be the norm are marked by significant habitat marginality and, in places, degradation, and a langur population density that is usually high. Environmental factors such as food quality, abundance and distribution, habitat destruction, predation, and the impact of humans (and their dogs) have been identified as important in understanding the pressures on these populations (Bishop et al., 1981). In some other locations where the animals were recognized individually, population densities were low, and where the habitat was good, for example at Orcha (Jay, 1963), Malemchi (Bishop, 1975), Junbesi (Curtin, 1977), and Rajaji (Laws and Vonder Haar Laws, 1984), no deaths were the result of males moving from group to group. One fact emerges from these many studies in different parts of India: langurs are capable of living in a number of kinds of social groupings with different interaction patterns and degrees of aggression according to factors of the environment, including population density and habitat quality.

Another very important set of factors ignored in the "infanticide" literature concerns specific kinds of experiences and interaction patterns that occur normally in male social development. Jay reported that infant and juvenile males often approached and contacted or attempted to contact relaxed adult and subadult males when the latter walked or ran past the youngsters and in other contexts when there was tension in the group (Jay, 1963; and personal observation). The approach and then contact by male infants with subadult and adult males is observed regularly in the colony. Young langurs, in the field and in the colony, often monitor the adult male or males carefully. When anything either unusual or out of the ordinary happens, the young males may race up to the adult, squeal or screech, stare at them, attempt to touch, embrace or simply remain close to the adult. Normally this does not endanger the young male, because the adult is not acting aggressively and usually ignores the immature. Sometimes, however, when there is significant tension and the adult male is being aggressive with others in the group, the adult seems unaware of the nearness of the young animal and may run into or over the youngster, or, as the adult strikes another animal, the young langur may also be the receiver because the latter is in the way. Since the young males do not avoid the adults when there is fighting or harassment, a course of action we humans certainly would recommend, the youngsters sometimes place themselves in grave jeopardy. What seems to be a typical and safe approach followed by contact under relaxed normal situations may become a death warrant for the young langur when adults are fighting seriously.

In addition, infants approach adult females when there is excitement, and adult females also injure and most certainly must kill infants. Placing the blame for death solely on adult males is a leap of faith based on assumptions, not evidence, because killings have rarely been witnessed.

Although we acknowledge the great behavioral variability within many primate species (Dolhinow, 1972; Fedigan, 1982; Smuts et al., 1987), we have seldom focused on the detailed dynamics of the processes, including developmental ones that are responsible for variability. Perhaps part of the reason we have not is that the answers require generations of observation, not simply a year or two. The questions we want to frame and the answers to these questions lie in patterns of development. There is no one convenient segment of life history that allows us to identify factors important in the expression of the many adult behavioral phenotypes of concern — those, for example, of rivalry and cooperation. To fully appreciate the possibilities requires knowing the genetics and development of the behaviors considered. We need to know pleiotropic effects, allometry, linkage, and the myriad forms of developmental constraints affecting the individual; otherwise, we will certainly overlook an array of possibilities that are available to the animal. As we watched generations of male langurs, some of what we assume to be many significant paths of development became apparent. In turn, the appreciation of early variability in experience encouraged us first to recognize and then to view the range of adult behaviors as normally available alternatives that constitute part of the repertoire of adult choices. Instead of characterizing adult male life as an artifact of rank-ordered relations measured by competitive outcomes, consider that the much more important measure of their behavior lies in the cooperative and coordinated relations of day to day life.

And Our Hominid Ancestors?

Let us now consider what langurs may suggest about our hominid ancestors. Although our records of general social behaviors from a few locations in India (for example, Jodhpur) are excellent, it is rather surprising how little we know from field studies concerning the ontogeny or development of the social behaviors that structure male relations. Based on more than two decades of observations of langur males in a stable colony setting, we describe some of the complexity and variability of male relations, including critical elements of behavior, such as cooperation, affiliation, opportunism, and perceptiveness. The essential opportunism of each male appears to be far more powerful than the inertia of established relations, as we observed individuals cast off old alliances when new ones appeared to gain them more than they would have by continuing with long-term ties. When males live together they form alliances of association and, at times, support — a characteristic not unique to langurs. Although male-male associations appear to be less important when males live in larger mixed-sex groups (for review see Walters and Seyfarth, 1987), they are still a prominent aspect of male langur life. By cooperating in alliances, many (if not most) langur males stand to gain more than if each acts as an individual. Significantly, these dyads or alliances are not based solely or even mainly on age or degree of kinship — although these factors may affect the formation of some alliances.

Our Pliocene-Pleistocene hominid ancestors were undoubtedly at least as opportunistic and able to cooperate as a group of monkeys when they perceived it appropriate. The social lives of our ancestors included females and immatures and were far more complex than the single-sex langur monkey group we described. However, our hominid ancestors shared with many living nonhuman primates, including langurs, similar life history structures. They all had to meet comparable challenges of daily life, including getting food and water, maintaining social relations, and keeping safe from predators. Many similarities in these patterns of behavior are based on a common phylogeny. Early (and late!) hominids and the Old World monkeys and apes are animals of similar physiological and morphological design and share common emotional-cognitive propensities. They most likely had, and have, similar responses to the problems of staying alive, reproducing, and rearing young. Many of these solutions must have been based on the ability to cooperate and recognize and act on responses that lead to mutual benefit, and many of these social networks took the form of special relations among males. These special relations, although at times apparently rather fragile and short-lived, can last for long periods and be exceedingly important to the individuals so connected. Our data underlined in red the caveat that once an ally, not necessarily always an ally! The individual male, in fact, does what he appears to deem best at the moment. This can include immediate dissolution of a relation that has been strong for months or perhaps years. Both cooperation and opportunism are surely part of the psychological armamentarium of the order Primates, and this is true for both sexes.

> It is not the fossil bones as such which are important, but the clues they give about the populations long extinct. These reconstructions contain very large subjective elements, and, if

we regard evolution as a game, then this fact must be considered, and we must be explicit as to the rules which are used
in the reconstructions (Washburn, 1973c:559).

Our guesses will be more informed if we base them not only on
the clues from the fossil and archeological record, but also on the
life patterns we and our close living nonhuman primate relatives
share. As our understanding of the behavior of modern primates
increases, we will overlook fewer of the possible attributes of our
ancestors. Our forebearers were probably not ruled by sex, age, or
kin, as has been assumed by too many authors (for example,
Ardrey, 1961; Lovejoy, 1981). We are an order characterized by
relatively high intelligence, and in the normal course of living we
respond to differences and change with attention and curiosity.
Behavioral variability (including all the mistakes from which we
profit), cooperation, and opportunism must have been central
themes in our early success story. Lastly, the evolutionarily recent
brain structures that form the neurological base for human
language made possible a vast amount of cultural elaboration based
on our experiences. The essence of modern humanity comes from
those elaborations.

II

TIME

The theory of evolution is not being weakened by this correction of past errors and misconceptions and by the opening of new understandings. On the contrary, the basic truths developed by Darwin and the brilliant succession of evolutionists who came after him are strengthened.

Science can make again for its goal: the unattainable but approximate truth about man, his origins and evolution.

—Ruth Moore (1953).

One of the most persistent arguments against the theory of biological evolution is that enough time, in terms of the earth's history, has not elapsed to allow for the major changes that have taken place. Given how different the facts appear today than they did even a few years ago, it is appropriate to discuss time and earth history: how differently science calculates time today than in the past and how challenging it has been to account for the vastness of the planet's history. The earlier, limiting ideas of time made it extraordinarily difficult for scientists to accept the theory of evolution. If the earth had only been in existence for a geologically short period, then the amount of time necessary for evolutionary change to occur remained inexplicable. Today, the theory of evolution is readily embraced by the scientific community as Tom Jukes discusses in his chapter entitled "Evolution." This has occurred in large part due to a more sophisticated understanding of the earth's history and geological time.

Some of the earliest assessments of time were based on an interpretation of the Scripture. For example, during the fourth century A.D., Eusebius of Caesarea (ca.260–ca.340) devised a chronology of the earth's history based on Jewish historical tradition. Later, Eusebius Hieronymus (Saint Jerome; ca.340–ca.420) extended the chronology back to Adam. Efforts to derive a history of the earth and its inhabitants from biblical sources culminated in the seventeenth century with the work of James Ussher. Archbishop Ussher did a careful study based on the Old Testament and astronomical cycles and calculated that creation occurred on October 22, 4004 B.C., a date which has since been printed in many editions of the Bible. John Lightfoot, a biblical scholar of the same day, calculated more specifically that the origin of man occurred in 3928 B.C. at 9:00 A.M. (Brice, 1982). Taken together, these dates provide the most specific calculations ever given for the creation event! In that day, even a 4000-year span prior to the Christian era was considered a remarkably long time.

The first formalized evolutionary system was published in 1749 by French scientist Georges Buffon (1707–1788). He estimated that the earth had an antiquity of approximately 75,000 years (Dean, 1981). He proposed that the earth started as an incandescent gas that consolidated into a molten ball over a 35,000-year span. He next reasoned that water vapors condensed and formed an uninterrupted ocean which covered the whole planet. Even his conservative estimate of 75,000 years was perceived as radical by the church and general public. To avoid censure, Buffon declared that his estimate was nothing more than "pure philosophical speculation" (Wicander and Monroe, 1989:23).

Two fundamental geological principles that emerged out of the scientific revolution of the 1700s profoundly influenced the development of nineteenth century geologists' understanding of time. The first is the *principle of superposition*, by Danish anatomist Nicholas Steno (1638–1686), which stated that any stratum is younger than the one it rests upon and is older than those above it (Geikie, 1905). Steno proposed that "fossils were once living marine plants or animals whose bodies were replaced by minerals," rather than quirks of nature (Thompson, 1988:44).

The second principle that greatly impacted scientists' understanding of the age of the earth was the *principle of faunal succession*. Independent studies in England (William Smith, 1769–1839) and France (Alexandre Brongniart, 1770–1847) established that "fossil assemblages succeed one another through time in a regular and determinable order" (Wicander and Monroe, 1989:63). From this, different eras of the earth's history could be consistently identified by their fossil content and the progression from simple to complex forms of life was illustrated. Thus, fossils became indicators of time.

Charles Lyell (1797–1875), following James Hutton's (1726–1797) theory of uniformitarianism, argued that all of the earth's land had at some point been submerged under the sea. He based this on the abundance of seashells deposited and fossilized in elevated continental strata. From his work, individual strata were grouped into formations and these formations were combined into higher categories which led up to the geological systems that we recognize today.

In 1852, Lord Kelvin (then William Thomson Kelvin) presented a paper titled "On the Universal Tendency in Nature to the Dissipation of Mechanical Energy." In this he stated that "in any conversion of energy from one form to another, a proportion of that energy will be converted irretrievably into heat" (Burchfield, 1975:13) and outlined what is today known as the second law of thermodynamics. At the time Kelvin presented his paper, he knew it had important implications for constructing the duration of geological time. An estimate could be derived by measuring the rate at which energy dissipates as heat. He reasoned that the earth was originally an inert molten mass, which progressively cooled at a constant rate over time. He calculated "the amount of energy available to the earth and sun from all conceivable sources" and then "determin[ed] the time necessary for each body to cool from its assumed primitive state to its present condition" (Burchfield, 1975:13–14). Kelvin concluded that evolution was not possible, because 25 million years divided by existing geologic strata does not yield enough time for the evolution of all life forms to occur.

In 1895, the year that Kelvin's calculations came into question, Wilhelm Conrad Rontgen discovered x-rays and set in motion a chain of events which revolutionized physics and medicine and also expanded concepts of geologic time. The following year, the French physicist Henri Becquerel (1852–1908) announced that compounds containing the element uranium spontaneously emitted energy similar to those exhibited by x-rays (1896). The geological significance of Becquerel's work was not acknowledged, however, until 1903 when Pierre Curie (1859–1906) and his assistant Albert Laborde "announced the discovery that radium salts constantly release heat" (Burchfield, 1975:163). In other words, they discovered what Kelvin had referred to as "the unknown source of energy." They announced the presence of two new elements— radium and polonium—and coined the term *radioactivity.*

Continued investigations into radioactivity raised the possibility that radioactive minerals might serve not only as sources of heat but also as timepieces to date the rocks which contain them. From this idea, the fundamental principles of radioactive dating were formulated. The concept of radioactive dating is relatively simple, although the actual procedures are not always so. In this section, the individual who pioneered potassium-argon dating (a type of radioactive dating), Garniss Curtis, discusses the contribution this technique has made to better our understanding of earth history.

With the development of radiometric dating techniques in the twentieth century, we enter into the era of modern science and with it the possibility of counting tremendous periods of time in objective ways. The conservative estimates at the end of the nineteenth century of 30 to 40 million years for the earth's age have soared to 4.5 billion years. The age of the mammals has gone from 3 million years to 70 million years. And the Pleistocene, the major period for the evolution of hominids, has gone from 200,000 years, as suggested by Sir Arthur Keith in 1931, to two million years.

Today, paleoanthropology can avail itself to many new radiometric dating techniques (though some have provided quite controversial results). Thermoluminescence (TL) and electron spin resonance (ESR) are two more recent dating techniques used in the field of paleoanthropology that have provided dates within a range of 40,000 to 100,000 years. Given that the emergence of modern *Homo sapiens* occurred during this time span, TL and ESR dating methods are providing critical information in dating this event.

TL dating draws on and measures the amount of trapped nuclear energy within a previously burned or fired artifact. All natural materials utilized for the manufacture of an artifact have a "geological radiation history" (Fleming, 1979:2). However, when the artifact is fired, as happens with pottery, or burnt for intentional or unintentional reasons, the naturally occurring radiation within the microscopic lattice of the material is erased. Thus, from the initial firing or burning, the artifact is set at its "time-zero point." During the heating process, a small amount of electrons are "knocked away from their parent-ion sites" (Fleming, 1979:1), and a small percentage of these go on to find a defect in the lattice of the material where they then reside. As time passes, energy builds and remains trapped in these "luminescence centers." When the artifact is then reheated in the laboratory, this stored energy is released in a thermoluminescence flash. The intensity of the flash is measured, and from this data the amount of time that elapsed from the first firing or burning to the present heating is calculated. Testing procedures can be carried out on ceramics, tile, bronze, burnt rock crystals, burnt stones, burnt flints, brickwork, and fabric (Fleming, 1979).

Electron spin resonance has been applied to the dating of tooth enamel. The mechanics of ESR dating are quite complex but in general the procedure detects

> unpaired electrons that have been trapped in a crystal lattice
> as a result of irradiation of the sample by alpha, beta, or gamma
> rays emitted from radioactive elements (U, Th, K) in the crystal
> and its environment (Grun et al., 1987:1022).

The age of a sample such as tooth enamel is calculated by determining the ratio between the accumulated dose of radiation in the enamel and the amount of radiation generated by uranium, thorium, and potassium in the surrounding soil. Calculations of both an early uptake of uranium (this assumes uptake occurred soon after burial) and a linear uptake of uranium (this assumes uptake occurred over time at a constant rate) must be made. The early uptake averages reveal the minimum possible age of a specimen, while the linear uptake averages generally correspond better with other dating method analyses (Schwarcz et al., 1988:736; Stringer et al., 1989:757).

ESR dating is not without its shortcomings. The impact of

"chemical reactions, heat, pressure, shock, and light" can reduce some ESR signals and thus produce erroneous results (Geyh and Schleicher, 1990:279). To avoid these potential problems, the history of the sample must be known so that appropriate adjustments can be made. Also, as with TL dating, loss or gain of uranium or thorium during the aging process can yield ages that are either too low or too high. A third problem with ESR dating is that water content "strongly influences the dose rate," and no way exists to determine the degree of water content fluctuation over the life span of a sample (Geyh and Schleicher, 1990:280). On the other hand, it has a very wide dating range, "from several centuries up to many hundred thousand years" and "analyses may be repeated many times with the same sample" (Geyh and Schleicher, 1990:273, 278).

It is with the dating of events surrounding the emergence of modern *Homo sapiens* that both TL and ESR have provided interesting, if not controversial, results. For example, excavations at Mount Carmel in Israel have yielded considerable fossil remains of both Neanderthal and early modern *sapiens*. Qafzeh and Skhul have produced numerous samples of modern *Homo*, while the geographically close sites of Kebara and Tabun have yielded distinct Neanderthal remains. Perhaps somewhat biased by the European Neanderthal sequences and a traditional belief that modern *Homo* emerged after Neanderthal, relative dating assigned 40,000 years to the *Homo* remains and 50,000 years to the Neanderthal material. These dates supported early contentions that modern *Homo* descended from ancestral Neanderthal stock and thus shared a common phylogenetic line.

In 1987, Neanderthal remains from Kebara were tested using the thermoluminescence method (Valladas et al., 1987). The results were not too surprising in that they confirmed estimates of approximately 50,000 to 60,000 years. The real jolt, however, came in 1988 when the *Homo* remains at Qafzeh were tested by TL methods (Valladas et al., 1988). These results produced a 90,000 to 100,000 B.P. (years "before present") date, pushing the date of the modern *Homo sapiens* remains back some 50,000 years. The following year, the anatomically similar remains from Skhul were tested using the ESR method (Stringer et al., 1989). Like the Qafzeh results, these hominids were dated in the 81,000 to 101,000 kyr B.P. (thousands of years before present) range. Further analyses carried out by other researchers using TL and ESR dating have verified these results (Schwarcz et al., 1988). From an independent line of evidence based on her work with mitochondrial DNA, Rebecca Cann also concludes that modern *Homo sapiens* arose independently of the Neanderthal line (chapter 5 in this section).

"Time is of the essence" is certainly a phrase that rings true when applied to the subject of evolution. Without vast amounts of time, Darwin's mechanism of natural selection could not be accounted for to the degree he proposed. The process of natural selection could be applied to the relatively short-term changes seen in the beaks of a few Galapagos finches or in the coloration patterns of some English moths, but the problem of speciation could have never been resolved. The Curies never could have known how significant their discoveries of radioactivity have been in solving one of Darwin's most significant dilemmas. There is still much work to be done in reconstructing our evolutionary pathways, let alone the evolution of the entire earth's history. As new dating methods emerge, new questions will be encountered and new issues raised. Such is the nature of scientific inquiry. However, now more than ever, with the knowledge of the operations of radioactivity, science has the means and resources available to reconcile problems and correct certain errors. This is all gratifying; forty years of Piltdown was enough.

Old Fossils and the Rocks That Dated Them

A Personal History on the Union of Geochronology and Paleoanthropology

Garniss H. Curtis

Garniss Curtis (Geochronology Lab, Institute of Human Origins, Berkeley) is recognized as the major contributor to potassium-argon dating techniques. This chapter reviews the history of this field and reports on the latest breakthroughs in technology which have further refined the results of this dating method. The name of this chapter was taken from the title created by Roderick A. McManigal for the Institute of Human Origin's Tenth Anniversary Symposium in 1991.

With the introduction of radiocarbon dating to the field of archaeology in the late 1940s, a new dimension was also added to the field of paleoanthropology. Although radiocarbon dating was restricted to events occurring in the last 50,000 years, owing to the short half-life of radioactive carbon, this range was sufficient to: (1) date events in the Mesolithic of Europe (2) show that Neanderthals were present there as late as 35,000 years ago, (3) show that *Homo sapiens sapiens* arrived on the scene in Western Europe almost at that same time. The timing and duration of events in the Paleolithic, however, remained speculative. Zeuner's great work on the Pleistocene epoch was the Bible for most paleoanthropologists. In that study, Zeuner argued persuasively for a duration of 600,000 years for the Pleistocene, and certainly no more than 1,000,000 years. This, of course, set a limit of the Paleolithic of Europe, for stone tools had been found only in interglacial deposits or, outside the limits of glaciation, in deposits thought correlative with glaciations or interglacials. Thus, in East Africa, where the Leakeys had found so many bifaced tools at Olduvai Gorge, Olorgesailie, Kariandusi and elsewhere, they spoke of pluvials and interpluvials when deposits were dominated by gravels (pluvials) or by fine-grained sands and silts (interpluvials), and thought them to be related to the same climatic changes that had brought on the interglacial and glacial periods in the northern hemisphere. No thought had been given to the effects of tectonism on the coarseness of sedimentary deposits.

While the accurate timing of archeological events in the Neolithic was being established by radiocarbon dates during the years following World War II, other longer-lived radioactive elements were being investigated for use as geochronometers. Of these, and of paramount importance to the dating of paleoanthropological occurrences, was radioactive potassium of mass 40 (40K). This radioactive nuclide decays to argon of mass 40 (40Ar) with a half-life of 1.25 billion years, which may seem shockingly large to the casual reader when it is compared to the half-life of carbon of mass 14, namely 5,730 years — a factor difference of 200,000 times. In actuality, it is much more than that! 40K also decays to 40Ca, and to our chagrin we have learned that a major amount of 40Ca is produced by this decay: in fact, 8.5 times as much 40Ca as 40K. This being so, one might ask, why don't we use the decay scheme 40K/40Ca for a geochronometer? The problem with this scheme is that calcium is a ubiquitous element in almost all minerals, thus some 40Ca is present when the mineral first crystallizes, to which is added 40Ca from 40K decay. It is difficult to correct for this accurately, although it can be done in very old minerals where the amount of initial 40Ca is relatively small. 40Ar, on the other hand is a gas, an inert gas, and most of the time it is absent from minerals when they first form. Since it does not have to be corrected for, greatly improved accuracy of ages can be obtained by this method.

We are still not out of the woods with respect to a major hurdle in this method of dating, however, because only one part of 40K occurs in approximately 8,500 parts of common potassium, the latter being principally potassium mass 39 (39K). This means that in order to obtain 40K/40Ar ages overlapping those obtainable by 14C, say, 50,000 years, one part in 500 million must be measured with high precision by the 40K/40Ar method. Surprisingly, the resolution of modern mass spectrometers is such that this can be done. We have obtained an age for the Campanian Tuff from the Bay of Naples, Italy, of 36,500 years +/– 400 years. This date is probably superior to the radiocarbon date of 28,000 years for this tuff owing to differences in the magnetic field of the earth at that time, and to a greatly reduced content of carbon dioxide in the atmosphere during the last ice age when this tuff was erupted.

These two factors change some of the basic assumptions of the 14C method. For example, during this time, and for reasons we do not fully understand, the carbon dioxide in the atmosphere was reduced 20 percent. We know this by measuring the carbon dioxide in air bubbles trapped in ice in Antarctica and Greenland ice caps. This means, then, that the amount of carbon dioxide formed from 14C during this ice age was relatively increased, because the cosmic ray flux striking the earth remained either the same or increased during this period of time. If the relative amount of 14C increased, it implies that 14C dates in this time span would be too young.

* * * * *

It seems inappropriate in the context of this volume to give a detailed history of the development of the 40K/40Ar method. For this history, one should read the superb and witty account by Houtermans (1966). For the history of the development of the 40Ar/39Ar method, a later modification of what is now referred to as the "conventional" K-Ar method, the book by McDougall and Harrison (1988) on 40Ar/39Ar dating is excellent. Because of my involvement in the early applications of K-Ar dating to paleoanthropological finds in Italy and East Africa, I shall give a personal account of how Jack Evernden and I, in the Department of Geology and Geophysics at the University of California, Berkeley, initially became involved in working with paleoanthropologists, a question frequently asked of me.

To begin with, to say that Evernden and I fell into potassium-argon dating, is not far from the truth. John H. Reynolds, a Ph.D. from Chicago in physics, joined the Physics Department staff at U.C. Berkeley in 1951. His initial objective was to design and make a mass spectrometer for studying rare gases in rocks and meteorites. By mid-1952, he had achieved this goal, and he came to me to get some geologic help in applying his new instrument—a mass spectrometer that was one hundred times more sensitive than any other such instrument in the world—to the dating of rocks using the decay scheme of potassium mass 40 going to argon mass 40. At that same time, I was making plans to go to Alaska for the purpose of studying Mount Katmai, a volcano that had erupted in 1912. It was a project I felt I could not drop, so I asked Robert Folinsbee, a visiting geologist from Alberta, if he would be interested in helping Reynolds. During 1953, then, while I was in Alaska, Folinsbee and Reynolds made great progress. However, owing to family problems, Folinsbee had to return home suddenly. He implored me to drop my Katmai project and take over working with Reynolds: "This potassium-argon dating system has an unbelievable potential. It's going to resolve all kinds of problems in geology and paleontology. You simply have to get into it, Garniss!"

Katmai, however, was a siren to me, and instead, I went to my close friend and colleague, Jack Evernden, who I knew was unhappy with his position in seismology under Byerly. My pitch to Evernden was hardly out of my mouth before he was out the door to talk to Reynolds and to begin our long and productive collaboration with him. I spent the field season of 1954 at Katmai. But shortly after I returned, Evernden came to me for help in separating minerals for K-Ar dating; gradually I, too, became totally involved in this new dating method. It was apparent to both of us that the great contributions of this dating method would be made in the younger part of the geologic time scale, namely the Cenozoic Period; and I felt sure that of the Cenozoic epochs, the Pleistocene epoch had most to offer.

Evernden and I had similar backgrounds in our undergraduate training, both graduating in the College of Mining at the University of California, Berkeley. This meant two years each of math, physics and chemistry as well as courses in engineering and geology. I liked the geology, Evernden the math and physics; so when we returned to Berkeley for graduate work, Evernden chose seismology and I geology, emphasizing volcanology. I had had a minor in paleontology as an undergraduate and had worked after graduation as a mining engineer for three years and as an oil geologist for one year, so I had a broad background in fieldwork before returning to Berkeley. It was during these four years that I began to realize that the Pleistocene epoch had to be longer than the 600,000 to 1,000,000 years assigned to it by Zeuner in his classic work — too much had happened to be encompassed in that short a period of time! Four major glaciations in North America and Europe struck me as being too few: the evidence for the older ones was so often nearly eradicated, how could we be certain we weren't confusing several older glaciations to which we were assigning one name? Moreover, there is a volcanic center in the Sacramento Valley called "the Sutter Buttes." These buttes, volcanic domes, intrude sedimentary strata ranging in age from late Cretaceous to late Pliocene, therefore must be of Pleistocene age. These more than five thousand feet of strata were turned up on end during this period of volcanism and eroded down to low hills by the end of volcanic activity, following which, the volcano itself was deeply denuded. Surely, I thought, this took more than one million years to accomplish!

Thus, after Evernden and I slowly got the bugs out of our K-Ar system, and the background of contaminating gases sufficiently low so that they didn't mask the small amounts of radiogenic argon we were attempting to measure, I looked at the Sutter Buttes rocks. Most of them were ideal for dating, for they contained either the high potassium-bearing mineral, biotite, or the even higher potassium-bearing mineral, sanidine; some of them contained both minerals. Our first try was on biotite from the very last eruption that occurred at the Buttes. In 1956, it took me two weeks to separate and purify the thirty grams we thought we needed for the experiment, and it took another three full days to make the gas extraction and mass-spectrometric measurement. With our present-day system, using the argon-40/argon-39 method, we would get a more accurate and precise date measuring the ages of five individual grains of biotite, each weighing less than one-thousandth of a gram. It would take us no more than an hour to separate and clean the grains, and then, less than three more hours to make the five age determinations! But for me, the 900,000-year age we obtained was worth all the time and effort. Later, we dated one of the older intrusions and found it to be 1.6 million years old: the Pleistocene was indeed old!

One of our objectives from the beginning was to obtain accurate calibration points for the geologic time scale on which the great Arthur Holmes had made a substantial start before World War II. I felt we were now ready to apply to the newly founded Miller Institute for Basic Research in Science on the Berkeley campus for funds for this project—funds which would not only give us travel money, but money to replace us with visiting professors for our courses. Evernden, for once unusually cautious, objected, but I dragged him, protesting all the way, to meet with Glen Seaborg, the reviewer of proposals of a physical nature for the institute. Seaborg was enthusiastic, and within a few months I was losing the toss of a coin to see who would spend the first year collecting samples in Europe from the type localities of the various eras, periods, and epochs that had been established there. It was well along into that first season that Evernden reached Rome and met Alberto Carlo Blanc, professor at the University of Rome and owner and chief editor of the journal *Quaternaria*. This was our introduction to the field of paleoanthropology. Near Rome, Blanc showed Evernden the Acheulean tool-bearing site of Torre in Pietra, which contained, in addition to numerous hand axes, a mid-Pleistocene fauna. Later, our K-Ar date of approximately 440,000 years from the sanidine-bearing pumice at the site was viewed with great skepticism by archeologists and paleoanthropologists, but not nearly the skepticism and roars of criticism that came from the results of the last part of Evernden's first field season!

Blanc introduced Evernden to L. S. B. Leakey in Nairobi, Kenya, by phone, and within a few days, Leakey was taking Evernden to

Olduvai Gorge. This was almost two years before Mary Leakey's
discovery of *Zinjanthropus* in Bed I of the famous site. At the time
of her discovery in 1959, we had already determined the age of the
lower part of Bed I to be 1.7 million years (Ma); with new decay
constants this is 1.8 Ma (Bishop and Miller, 1972). We didn't know
the relationship of the dated tuff to the Zinj site, so I returned at
the first opportunity and made a more extensive collection of
samples that confirmed Evernden's original date, and which applied
also to the nearby and slightly older find of what was to be named
Homo habilis. In fact, we showed that all of Bed I was deposited
in probably less than 150,000 years. Bed II, however, with abundant
handaxes in its upper part, *Homo erectus* skull in its middle, and
Homo habilis remains in its lower part, could not be dated
accurately at that time because all of its volcanic tuffs were derived
and heavily contaminated. (Recently, using the carefully detailed
stratigraphy of Olduvai Gorge established by Richard Hay in his
monumental monograph on the geology of Olduvai Gorge [1976],
Robert Walter and colleagues at the Geochronology Center in the
Institute of Human Origins at Berkeley, together with Paul Manega,
a Tanzanian student at the University of Colorado, have redated
the tuffs at Olduvai using the single-grain argon-40/argon-39 laser
fusion technique. The results from Bed I have been published
[Walter et al., 1991] giving beautifully precise dates for several tuffs
in that unit, and modifying but largely confirming the work of thirty
years ago. They show that the duration of Bed I is even more
narrowly constricted than originally thought. The dating of Bed II
has been much more successful, as the dating of individual crystals
permits the elimination of contaminating older grains. This work
is near completion and will be described in Manega's forthcoming
Ph.D. thesis.)

Our first results at Olduvai were hardly greeted with open arms
by paleoanthropologists and archeologists, and, for that matter,
geologists; in fact, it was just the opposite! "Evernden and Curtis
have doubled if not tripled the length of the Pleistocene which
Zeuner showed conclusively is no more than 600,000 years. This
is highly improbable." (That from an anthropologist.) "There are
many reasons not to accept these dates. It is highly probable that
the crystals used for dating had argon in them at the time of
eruption, hence were old when they were erupted" (That from a
geologist.) Wrong! We had worried about just this problem in our
early experimentation, so had examined carefully all of the historic
eruptions that had suitable minerals for dating: a volcanic tuff from
the island of Ischia in the Bay of Naples, a 1954 eruption in New
Guinea, the 1912 eruption of Katmai in Alaska, and so on. At the
limit of our detection, +/− 5,000 years, we could detect no excess

radiogenic argon in these rocks. Five thousand or 10,000 years added to 1.8 million would not, of course, change the date appreciably. Others would show later that some rocks do indeed contain significant amounts of excess argon when they are erupted, particularly some lava flows that are chilled so quickly that their initial argon cannot escape. We suspected this even before it was proved; hence, we avoided using such rocks for dating.

While we were the first to apply the K-Ar method to dating early human remains and artifacts, other workers in both the United States and Europe were also beginning to apply the method to dating fossil-bearing rocks. Of these, Lipolt at Heidelberg attempted to show that our dates for Bed I, Olduvai, were too old. He obtained a piece of green basalt, a tool, I recall, from Bed II, thought to have been made from lava at the base of Bed I, for which he got a K-Ar date of approximately 1.3 Ma. He argued from this that our dates must be wrong. An advantage that Evernden and I had over most physicists doing K-Ar dating (and that was most of the people at that time working with K-Ar), was our extensive training and familiarity with the petrology of rocks. Most volcanic rocks that have been buried for any length of time, even shallowly buried, are altered chemically by ground water moving slowly through them. Such alteration causes loss of radiogenic argon which leads to K-Ar dates that are too young. Petrologists readily recognize such altered rocks and avoid them for dating purposes (although sometimes individual minerals can be separated from an altered groundmass and successfully used for dating). Clearly, this is what happened to the rock that Lipolt dated. Shortly after this annoyance, we got support for our dates from Bed I from a totally unexpected source.

An entirely new method of dating had been developed by three scientists at the General Electric Research Laboratory based on a rare form of decay of uranium. Normally, uranium decays by emitting alpha particles, ionized helium atoms of mass four. Occasionally, however, the uranium atom simply explodes— fissions—splitting apart into two pieces that shoot off in different directions with great energy. If the fissioning uranium atom is in a mineral, the two energetic fragments break the bonds of the elements composing the mineral that are in their paths. The tracks, the holes, made in this way are invisible to the unaided eye, but can be made visible if they occur near the surface of the mineral by etching with a suitable chemical reagent. Thus, if the etched tracks are counted under a microscope, and the total amount of uranium in the mineral is determined, and if the rate of fission decay of uranium is known, the age of the mineral can be determined. Price and Walker were the scientists who first invented this technique; it was they, together with Fleisher, who applied it to dating volcanic glass from Tuff IB of Bed I, Olduvai (Fleisher et al., 1965). This is the same member for which we had obtained an age of 1.8 Ma; they determined a fission track age of 2 Ma +/– 0.25 Ma. This essential concordance of dates by two entirely different decay schemes convinced all knowledgeable people that the human evolutionary line was much older than had heretofore been believed.

Surprisingly, though, as the Olduvai dates were slowly being accepted, the K-Ar dates for other archeological sites in Kenya that we ran at that same time, were almost totally rejected! Malawa Gorge was a tool-bearing site of Leakey's near Lake Naivasha that he correlated with the Stillbay culture of South Africa. The small bifaced obsidian tools found there are beautifully and distinctively made, and did indeed look to me like the illustrations he showed me of type Stillbay tools. "These can't be older than the last high level of Lake Naivasha during the last glacial period." Leakey assured me, "30,000 years at the most." The sample I collected for dating was a tuff rich with large euhedral sanidine crystals, lying only inches above a tool-bearing stratum with an obsidian tool in place protruding from the edge of the bank. Two K-Ar runs of this sample gave almost identical ages of approximately 240,000 years. "You're wrong this time, Garniss," Leakey exploded, "your dates can't be right, simply can't be right!" For the nearby Wetherill and Cartright sites high on the escarpment overlooking Lake Naivasha and containing somewhat more crudely crafted obsidian bifaces that Leakey termed "Pseudo Stillbay," my dates of approximately 350,000 years were again rejected. "There can't be more than a few thousand years separating these two quite similar cultures!" exclaimed Leakey. Desmond Clark agreed with Leakey about the age of the Stillbay culture: "I've just gotten a charcoal date for Stillbay of 60,000 years from DeVries in Holland. Stillbay can't be 240,000 years old!"

Kariandusi, where Leakey had established a small park to protect several large Acheulian-type hand axes, gave K-Ar dates of about 900,000 years which, again, were unacceptable to archeologists: "Either your Torre in Pietra dates are wrong or your Kariandusi dates are wrong. They both can't be right," said Robert Heizer at Berkeley. "It has been shown many times that the cultural evolution of man has been synchronous wherever he is found, and no matter how widely separated," he continued. "Oh," I replied, "how has that been proven?" "A tremendous amount of detailed work has been done on this problem," he explained, "It just can't be questioned."

By the time Glynn Isaac and I obtained a date of approximately

1.6 Ma for hand axes near Lake Natron, it had begun to dawn on archeologists that hand axes had been used by humans for a very long time. Some of our early dates, by this time, had been confirmed by other laboratories, such as Paul Damon's at the University of Arizona and John Miller's at Cambridge.

Miller must be credited with making the single most important contribution to the precise calibration of hominid evolution. He was the first to apply a method of dating to hominid sites in Kenya. The method had been developed by two of John Reynolds's students, Craig Merrihue and Granville Turner, at Berkeley in the early 1960s. This method is a modification of the K-Ar method and is called "the argon-argon method," or "the 40Ar/39Ar method." In conventional K-Ar dating, the two isotopes, K and Ar, are measured separately, a mass spectrometer being used for the argon measurement and a flame photometer generally being used for the potassium measurement. In 40Ar/39Ar dating, some of the potassium is converted to argon of mass 39 by first placing the sample in a beam of fast neutrons. The sample is then fused to release its argon, as in conventional K-Ar dating, but now, by measuring the 39Ar and the 40Ar at the same time on a mass spectrometer, both the potassium and the argon in the samples are determined simultaneously. Known standards, that have been irradiated with the unknowns, are used to monitor the results. This leads to much more precise age determinations than conventional K-Ar dates, and if done very carefully, to more accurate age determinations as well.

Miller's great contribution to the use of this method for dating early hominids and artifacts in East Africa must be explained. When a sample is irradiated with neutrons, elements other than potassium that occur in the mineral (calcium, for instance) yield some argon as well. The quantities of these additional argon isotopes are small, and in very old rocks such as meteorites or rocks from the moon, they hardly interfere at all. But with geologically young rocks with very little radiogenic argon, their presence, if uncorrected for, is disastrous. Miller recognized that if all of these contaminating isotopes could be accurately corrected for, the argon-argon system could be used to date young rocks. He put a graduate student, N. R. Brereton (who used the solution for his Ph.D. dissertation) to work on this problem (Bishop and Miller, 1972). Subsequently, Miller used this method to date the stratigraphy at Koobi Fora in East Turkana where so many important hominid fossils have been discovered. McDougall at Canberra, using this same method, has continued to make important contributions to the dating of fossils on both the east and west sides of Lake Turkana.

The next important step in argon dating was made by Derek York at the University of Toronto. When Oliver Shaeffer, who had been experimenting with laser fusion at Stony Brook, died, York took over using a laser beam to fuse minerals. One of the advantages of this method of fusing mineral grains to release their argon is that an intensely hot beam can be focused directly on a single small grain and melt it without heating the container holding it. Thus, unwanted gases, which adhere to the container walls and which interfere with the precise measurements of the radioactive argon, are not released. This permits using single crystal grains for obtaining dates. The advantage is this: volcanic tuffs that have been contaminated with older derived minerals can now have each of their grains accurately dated. Several grains, five to ten or more, are selected from the tuff and individually dated; contaminating older grains are immediately detected. All those grains giving the same and younger age are the true age of the tuff. The precision of dates obtainable by this method is truly remarkable.

We at the Geochronology Center at the Institute of Human Origins (IHO-GC), Berkeley, have been able to obtain dates from tuffs near the Cretaceous-Tertiary boundary of 66 Ma +/− 60,000 years! Our contribution to this technique has been to automate it completely. Approximately 145 individual grains are placed in small pits in a circular copper disk in an evacuated extraction line. Using a computer, we program a movable stage which holds the disk to move each grain to be fused, in turn, under a laser beam. The gases released are cleaned in an automated cleaning system, then are released through a programmed and automated valve to a mass spectrometer for measurement. The data are collected and processed by the computer, and the age automatically computed. The whole process, from the moment of fusion to the appearance of all of the data plus the date itself, takes a half hour or less. Thus, all of the grains can be dated in about three days without anyone having to be in attendance! We are now using this system to date samples from paleoanthropologic projects in progress at Hadar, the Middle Awash, and the Omo area in Ethiopia; Olduvai Gorge and Laetoli in Tanzania; and several areas near Lake Baringo in Kenya. The remarkable results that Deino and Potts achieved at Olorgesailie have already been published (1990) as have the previously mentioned Bed I, Olduvai dates (Walter et al., 1991).

Since the early applications of 40K/40Ar dating to paleoanthropologic finds, several other methods of dating have been introduced to this field with more or less success. Some of the these methods are based on radioactive elements, some on other rates of nonradioactive change that can be calibrated. The decay of 87Rb to 87Sr has long been known and applied to dating igneous rocks. Igneous rocks as young as 200,000 or 300,000 years have been dated with high precision. The problem of applying this method to ·

tuffs burying hominid artifacts or hominid remains lies in correcting for initial 87Sr in the sample. Several different primary minerals must be present in the tuff to obtain data necessary to correct for initial 87Sr, and in most cases, only a single mineral occurs with the altered igneous glass of the tuff, thus making corrections for initial 87Sr questionable. It should be noted, however, that there are many cases where this method could and should be applied to obtain dates concordant with 40Ar/39Ar dates.

Disequilibrium series dates are based on the fact that the two isotopes of uranium, 238U and 235U, and the one isotope of thorium, 232Th, decay at different rates but in a series of similar steps. Each of these elements decays by emitting alpha particles, helium of mass 4. After eight decay steps, 238U ends up as lead of mass 206, which is stable and does not decay. 235U decays in seven steps to stable lead of mass 207; and 232Th decays in six steps to form stable lead of mass 208. For each of these elements, each step takes a different period of time. Weathering and other geochemical processes may separate daughters of uranium or thorium decay from their parents. Thus, a member of a series separated from its parent decays at a rate determined by its half-life yielding a daughter which in turn decays at a rate determined by its half-life. The ratio of parent-to-daughter isotopes will change until equilibrium is attained. As long as the ratio of parent-daughter isotopes is changing, the ratio may be used as a chronometer, but once equilibrium is reached, the ratio stays fixed and can no longer be used for measuring age. Several different pairs of parent-daughter isotopes from uranium and thorium have been used to measure geologic time from a few years up to a million years or more. Critical for the successful application of these radioactive pairs to accurate dating is the assumption that during the decay of parent to daughter, there has been no loss or gain of parent or daughter but that they have remained in a closed chemical system. This is often a difficult assumption to prove.

Thermoluminescence dating has been increasingly used for determining the age of paleoanthropologic material. The principle is easily understood: ionizing radiation coming from alpha and beta particles released by decay of radioactive elements in the rock or mineral to be dated or from cosmic rays bombarding the mineral or rock may interact with elements therein and release electrons which become metastably trapped in defects in the crystal structure, thus retaining some of the energy they received from the bombarding rays or particles. Heating the rock or mineral causes the metastably trapped electrons to fall back to their stable ground states, emitting light as they do so. This light emitted by heating is called thermoluminescence. The amount of light emitted is a function of the time passed since ionization radiation started and the amount of radioactivity that has been present in the mineral or rock. Techniques have been devised for these measurements and the method has been applied to a wide variety of materials with considerable success. Crystal structural defects may become saturated with electrons if there is excess radiation, and this may limit the age obtained. Specimens also may become altered in a variety of ways causing them to lose their metastable energy.

Electron spin resonance (ESR) is one of the newer methods of dating and offers promise of directly dating some specimens that can be dated in no other way. The principle involved is very similar to that in thermoluminescence. Metastable electrons produced by ionizing radiation and trapped in structural defects may be detected by an ESR spectrometer. This instrument works on the principle that when the specimen is placed in a strong magnetic field the unpaired spinning metastable electrons are aligned either parallel or antiparallel to the applied magnetic field. Under these conditions the electrons will absorb energy in the microwave region at specific frequencies. The amount of energy they absorb is a function of their number, which, in turn, is a function of the age of the specimen and the amount of ionizing radiation it has received. Advantages over thermoluminescence are: (1) it is less sensitive to premature release of stored energy during sample preparation; (2) the stored electrons in structural traps can be measured without draining the traps, thus measurements can be repeated for increased precision; and (3) sample preparation can be made at room temperature. As with thermoluminescence dating, traps may become saturated with electrons, after which the method is inapplicable as the obtained age is that of the time of saturation.

When high-energy cosmic rays strike rocks at the surface of the earth, elements in their minerals are broken up and new elements produced. These are called spallation or cosmogenic elements, and they may be used to date the time the rock has been exposed to cosmic ray bombardment. Some of the elements so produced are 3He, 21Ne, 10Be, 26Al, and 36Cl. One sees immediately the potential for using these cosmogenic elements for dating ancient stone tools that have been lying on the surface of the ground, or for dating thin rock deposits burying such artifacts. Obviously, to be successfully dated, one must assume that the rock or tool has not been disturbed since it was deposited or that weathering has not removed the original exposed surface of the rock or artifact (Faure, 1986).

Volcanic glass or obsidian gradually adsorbs water and becomes hydrated. The rate of hydration is principally a function of the chemical composition of the glass and of the ambient temperature;

the warmer the climate, the more rapid the hydration. The line between hydrated and unhydrated glass is sharp and easily detected with a microscope. Volcanic glass tools can be accurately dated if both parameters, temperature and chemical composition, can be determined and thickness of hydrated layer accurately measured. Unfortunately, with very old tools part of the hydrated layer may have spalled off, or the glass tool may have become completely hydrated so that only a minimum age may be obtained.

Dating organic fossil material using amino acid racemization is based on the fact that all proteins in living organisms consist principally of one of two possible structural types — enantiomers — that are stereomers of each other, both being optically active. In living organisms, the L-enantiomer amino acids dominate to the virtual exclusion of their stereomers, the D-enantiomers amino acids. After death of the organism, the L-enantiomers undergo slow transformation to the D-enantiomers, the equal mixture of the two forms becoming optically inactive or racemid, hence the term racemization. In short, the D/L amino acids ratio increases with the age of the organic material and approaches the limit 1. The rate of racemization is largely a function of temperature, being almost zero at zero Celsius. To apply the method successfully, one must know, or be able to determine with high accuracy, the average temperature of the site of the specimen. Unfortunately, this is seldom possible with old specimens — older, say, than the Holocene (Curtis, 1981).

With this large choice of dating methods at the disposal of paleoanthropologists, there are few sites, indeed, of hominoid-hominid fossils or artifacts that cannot be approximately dated, and in many cases, very accurately dated. It should be emphasized that, wherever possible, more than one dating method should be applied to a site in order to establish accuracy through concordance of dates.

Genes, Dreams, and Visions
The Potential of
Molecular Anthropology
Rebecca L. Cann

Certainly the contributions of molecular biology have revolutionized the study of human phylogenetic relationships. Along with the late Allan Wilson, and Vincent Sarich, Rebecca Cann (University of Hawaii) has been at the forefront of applying molecular studies to problems in human evolution. Her most recent collaboration with Wilson (Scientific American, *April 1992) addressed the question of the origins of modern* Homo sapiens. *In the chapter provided here she reviews the contributions of molecular anthropology to these and other major phylogenetic issues of the day.*

The ancient prayer of the kahuna from the temple tower was, "Let that which is unknown become known." It was hoped that during his apprenticeship each priest would learn fast (a'apo), obtain great patience (aholoa) and strive to obtain understanding (na'i) that he might become a lover of wisdom and knowledge ('akeakami) and be able to peer into the depths of his profession as the fisherman can look into the sea when he has put kukui oil on the water (McBride, 1972:8-9).

Students of anthropology are some of the best observers of human frailty in the universe, aside from preschool teachers. They watch for patterns and inconsistencies in behavior, for the presence and absence of adaptations to environmental conditions, and for clues about the processes which take place in the brain that governs socially interacting bodies. This fascination with understanding humans, their unique primate heritage, and their place in nature continues unabated into the new century. Moreover, the study of primate and human evolution is increasingly enriched by nontraditional anthropological methods, especially those from molecular biology.

The purpose of this essay is to provide an overview to students about molecular approaches to the study of human evolution that have influenced and even transformed the way physical or biological anthropologists talk about our past. That which was unknown is becoming known, because an integrated organismal and molecular approach to the biology of humans is now possible. The focus employed here implicitly assumes that the structure of DNA is a living fossil, conveying as much evolutionary information as mineralized dead cells. It assumes that biochemical methods for extraction of informational macromolecules from fossilized materials will become more efficient, quicker, cheaper, and easier to perform. It builds on a body of data that has documented genetic diversity in populations caused by the relentless accumulation of mutations in DNA sequences over time. Finally, it is as reductionist or as holistic as the reader. Therein lies its power.

Genealogical Searches

Power, validation, and increased prestige by association with individuals of higher status has motivated humans to establish their biological descent from famous entities, be they gods or humans. The outcry raised against the hypothesis that all individuals alive today can trace some of their genes to a woman who lived in Africa approximately 200,000 years ago should be put in this context, as well as a paleontological one. Would this hypothesis originally have been greeted with such outright scorn if DNA polymorphisms were first used in anthropology to document the descent from William the Conqueror of white male faculty at elitist educational institutions, instead of describing their linkage to an unnamed female individual of probably low economic status?

History documents the illogical obsessions of groups to establish superiority over each other based on evidence of descent from more illustrious ancestry. Genetic "proof" of this superiority by genes in the nucleus, which form the basis of most heritable change, is impossible to achieve. Simply put, original parental combinations of genes are repeated exactly in offspring only at extremely rare probabilities. With 23 pairs of chromosomes per adult cell, the total chromosome combinations which could occur in any offspring are 2^{23}, or 8,388,608 possible karyotypes. Only two of these will be exactly the same combination of chromosomes as either original parent, or one in about 4.2 million, assuming no crossing over along the chromosome arms. Boyd (1950) pointed out that there is a small but distinct chance that not a *single* one of your chromosomes came from a given grandparent, one in 4,194,304 (I have adjusted his number to reflect that the haploid chromosome number of humans at that time was incorrectly thought to be 24). In fact, you would have a one in 300 chance of not getting more than even 5 of your 46 chromosomes from this favorite grandparent, again, assuming no genetic recombination. Boyd performed these calculations to demonstrate that in the unlikely event that one is able to directly trace one's ancestry back to a famous person like William the Conqueror, the chances that you are any *more* similar to him genetically than you are to a very large number of people alive at the same time is vanishingly remote. Descent even from a deity does not guarantee heritability of the godlike traits.

Mitochondrial DNA Breaks the "Rules"

However, what was also unknown in 1950 was that while most genes (or more accurately, their DNA sequences) of the cell are contained on the chromosomes found in the nucleus, a small number are found only in the mitochondria of the cytoplasm. The mitochondrial genes demonstrate uniparental instead of biparental inheritance, where some characteristic of the offspring of two parents is determined by a factor present in only one parent. Maternal inheritance of mitochondrial DNA (mtDNA) means that while males have mtDNA, they do not normally contribute it to their offspring. General reviews of mtDNA and its importance in stimulating new directions in evolutionary biology are abundant.

Mitochondrial genes control energy production and are absolutely essential to cell function, but they do not undergo sexual recombination, segregation, or independent assortment like genes in the nucleus. So, it is possible to demonstrate with mtDNA that individual humans today share unbroken maternal genetic lineages uniting them with their distantly more remote female ancestors. The DNA sequences of these genes change over time, overall at about 2-4 percent per million years, but mutation does not obscure our ability to connect ancestors and their living descendants. If we count the number of differences in DNA sequence between modern people, and know how fast the differences accumulate, we can tell when two people must have last shared a common ancestor which contained the primordial mtDNA sequence. Thus, mtDNA polymorphisms and the concept of a molecular clock can be used to infer evolutionary change in modern populations.

Our inferred African Lucky Mother of all individuals alive today was not our only genetic ancestor, nor was she the only woman alive at the time who may have contributed genes to modern people (Cann et al., 1987; Vigilant et al., 1991). However, she did found the lineage of mothers who connect us in an unbroken link to the past. She may be the only common ancestor we all share, and for that reason, she has been called a mitochondrial "Eve" by some. You are probably no more like her in total genetic makeup (some 100,000 or so genes) than you would be to any other person living at that time, but for your 37 mitochondrial genes, she is your lifeline. Population geneticists estimate that she might have belonged to a group of 9,000-18,000 individuals (Hartl and Clark, 1989). Biological and cultural characteristics of her social group are therefore of broad interest to anthropologists, because we would like to know what other similarities between them and us might be present.

Methods That Drive the New Analysis of Human Evolution

Genetic characteristics of ancient and modern populations are being revealed with a technique called polymerase chain reaction (PCR). Genes are made of DNA, and the first step in understanding how genes work is to decipher the linear sequence of subunits, or bases that make up the gene. This technique builds on biochemical knowledge of how nucleic acid subunits, the DNA bases adenine (A), guanine (G), cytosine (C), and thymine (T) are joined together in a long polymer by the enzyme DNA Polymerase (DNA Pol) in the presence of the appropriate salts and buffer (Erlich, 1989). If a partial sequence for a piece of DNA is already known, it is possible to make a short stretch of synthetic DNA, say 18-20 subunits, that will serve as the start site where new polymerization will begin. The synthetic DNA is called the primer, because it primes new synthesis of the growing chain. If another stretch of sequence further along the DNA strand is known, a second primer, facing in the opposite direction, will send synthesis of new DNA back in the original direction. If the molecule is heated, all strands dissociate. When it is cooled, primers will anneal back to the strands—both the newly made ones and the original copies. This reaction can be repeated. The number of new strands made increases exponentially, and, if the reaction is repeated 25 or 30 times in the same test tube, a phenomenal number of strands are formed.

Molecular geneticists exploit this technique to make millions of copies of any DNA piece without having to grow bacteria, splice genes, or screen thousands of potential clones from a gene library that might contain the one DNA sequence of interest in a background of several million others. It is quick, in that a single afternoon in the laboratory with a single hair can give one enough DNA to perform a week of DNA sequencing, base by base. Innovations exist for what one does if just enough information is available to construct a single primer, or if one wants to amplify ten different sequences all at once in the same reaction vial.

The advantages of this technique, over conventional cloning using bacterial or yeast cells, are many. Cloning actually produces mistakes in the DNA sequences detected, because the bacterial replication of animal DNA is sloppy. Successful cloning also requires very pure, biochemically intact DNA. Cloning is often time consuming, because of the preparation of vectors and the growing of host cells. It is also somewhat of a long shot, if what one wants is a relatively rare sequence that may exist only in a limited population of cells from a certain developmental stage.

PCR allows scientists to increase their resolution, shorten the time of their analysis, and eliminate many costly steps that restrict DNA sequencing to fresh cells from a representative few individuals in a breeding population. In the past, studies of genetic polymorphism at the DNA level required heroic efforts. Even now, it is too costly without PCR. Indirect assessments of population variability using restriction enzymes and purified or cloned DNAs are now being supplanted by direct studies of the same genes from single hairs or feathers, tiny amounts of blood, 0.5 grams of powdered bone, or 3 millimeter slices of dried skin using PCR.

PCR, mtDNA, and Human Evolution

It was important to apply PCR to the question of modern human origins, because a number of problems could have complicated the interpretation of the original study using restriction enzymes and indirect techniques for counting differences between modern human mtDNAs. First, the question of the appropriateness of substituting Afro-Americans for authentic African donors was raised. Second, many researchers disagreed that a molecular clock had been firmly established, and argued that the time inference was incorrect. Third, others felt that the number of people originally examined was too small (less than 150) to make global pronouncements. Additionally, the phylogenetic method for constructing the human tree may have been inappropriate, or assumptions about the way in which the ancestor was inferred may have been inappropriate.

In the last five years, many laboratories have become engaged in human population studies using mtDNA. Three new studies are mentioned here, which confirm the original study and enlarge our understanding of evolutionary processes. It is now possible to have greater confidence in the African origin of modern humans from DNA data alone, because chimpanzee mtDNA sequences confirm that relative to the great apes, the most divergent human mtDNA sequences known are found among Africans (Kocher and Wilson, 1991; Vigilant et al., 1991). Outgroup rooting with chimpanzees demonstrated the African radiation basal to the gene pool of all modern peoples. Divergence implies great age, because mutations take time to accumulate in a population. Unless one wants to argue that for reasons unknown to molecular biologists, Africans suffer higher mutation rates, this result could only be obtained if Africa contains the ancestral population from which we all arose.

This greatly enlarged study of human mtDNA evolution, using hypervariable mtDNA sequences donated by ethnic Africans from a number of tribal groups, more Asian donors from isolated geographic areas, and additional donors from the Middle East, has confirmed the great diversity of the !Kung and forced a recognition that all other geographic isolates are really just a maternal kin group of African aboriginal peoples. The probability that a non-African group will lie at the base of the human mitochondrial tree is exceedingly small, because the first fourteen clades stemming from the mitochondrial mother are exclusively African (Vigilant et al., 1991). This finding magnifies the problem of accounting for the loss of Asian *Homo erectus* lineages, if these taxa are really directly ancestral to modern people from Asia.

One important new perspective about population dispersal and colonization of the Americas has also recently emerged, stemming from PCR studies of tribal groups in the Pacific Northwest. Four-deep maternal lines, with a last common ancestor estimated to have lived over 60,000 years ago, have been detected in a single tribe of Native American donors (Ward et al., 1991). If Beringia represented a significant barrier to human dispersal, with population bottlenecks determining the level of genetic diversity for millennia afterwards, this finding is unexpected. Either many more people took part in the trek across the Bering Sea than had previously been estimated from low resolution mtDNA studies (Schurr et al., 1990), or colonization of the Americas took place at least three times as long ago as most archaeologists and linguists are currently willing to accept. Either way, new questions are raised about issues that seemed previously understood.

Diachronic Comparisons for Population Continuity

New anthropological science focused on the study of ancient DNA is possible now, because PCR can amplify the one intact DNA sequence present in a bone sample that is only partially mineralized or soft tissues that are mummified (desiccated). Five-thousand-year-old human mtDNA sequences are common under some conditions, although most of the authentic DNA obtained from these sources is from stretches only 100-200 bases long and does not reflect the entire base sequence of whole genes (Paabo et al., 1989). Some nuclear gene sequences, usually from repetitive gene families, have also been reported from ancient materials.

If one wishes to talk about the effects of natural selection, or, sexual selection, on the survival and expansion of traits in a population, the linkage of DNA sequences to those traits and the identification of DNA in fossil populations could strongly support or refute hypotheses about human evolution. DNA sequences from the before and after stages of an episode of natural selection or a dispersal event could tell us much about the evolutionary process that is unavailable to paleontologists lacking a time machine.

The ever-present threat of modern contamination of ancient DNA, however, requires scientists to separate the preamplification and postamplification analyses of their materials into different rooms. DNA saturates laboratory walls, air ducts, equipment, buffers, and lab coats. In a laboratory where human DNA is routinely screened, it may be necessary to collect voucher hairs from visitors and resident workers alike for comparison to potential experimental results. Researchers of European descent must be especially careful in performing studies on European neolithic DNAs.

Proof of successful amplification of ancient DNA also requires information about that sequence, not just the observation of a fluorescent band of DNA on an electrophoretic gel, with a size corresponding to that which is expected experimentally from the placement of the two primer sequences. Short fragments of DNA amplified in early cycles can also recombine to form chimeric sequences which were not present in the original starting material. Practitioners of PCR approaches on ancient materials are currently in the process of sorting out the rules of conduct, the necessary demonstrations of rigor, and the levels of evidence essential to demonstrate the survival and identity of ancient DNA sequences (Eglinton and Curry, 1991). The possibilities raised in Jurrasic Park are not yet in hand and researchers have been sensitized to possible hazards, as well as advantages, that this approach to genetic investigation may entail.

Potential for Future Research

Biological anthropologists often feel that their work is free of some of the biases present in other fields of anthropology, but this is clearly not the case (Harraway, 1991). It is possible that within biological anthropology, some fields of study may be more objective in their analysis of potential evidence, their judgment of what constitutes evidence, and their framework of what problems are even worth studying. However, speculation impassions researchers to continue research long after hard facts have been filed away in the conscious brain, and speculation in one subfield can often serve as a force to test related hypotheses from a different perspective. The survival of Neanderthal mtDNAs in modern Europeans is one just prediction stemming from regional continuity hypotheses (Mellars and Stringer, 1989). When extraction methods are perfected, Neanderthal skeletal collections will become increasingly more valuable research materials, perhaps even more valuable to molecular biologists than the rarer Australopithecine collections.

DNA sequences do stand for all time, if they can be detected and deciphered. They are a universal code, but the rules by which they conduct development are sometimes unclear. In this regard, they can share a drawback in common with fossil evidence: proof that some form existed with a suite of characters, but the relevance of that form and those characters to present populations must be inferred from additional data. At the same time, DNA sequences can be compared between extant and skeletal populations for estimates of survivorship and extinction. They can also be directly correlated with agents of selection, such as infectious disease or human aggressors. If the appropriate techniques are discovered that allow the extraction of DNA from disease-producing micro-organisms preserved in skeletal remains, a part of the whole human and microbial ecosystem at the time of death will be open for study. Genotypes surviving over time in human remains will be followed actually at the DNA sequence level, not just modeled by computer simulations.

Perhaps it is too ambitious to hope that a geographic grid could ever be laid on the habitable surface of the planet and that human DNA donors could be recruited every five kilometers for a global study of migration, mate choice, and morphological change. However, the results of such a study would probably astonish us with the detail of information about human populations and forces of evolution that have resulted in anatomically modern people.

For contemporary United States populations, based on data from high school reunion booklets, migration is farthest in the stage of birth to graduation from high school (Koenig, 1988). There are sex differences and education differences that are detected regarding dispersal distances, marriage, and birth of children, with females marrying closer to home than males, but also dispersing further between high school and current residences than males. What new cycles of migration are changing the genetic population structure of North America?

Females and their uniquely transmitted genes may be actually better markers of migration fronts than males, although it is commonly assumed that the burden of offspring, their importance as a biological resource, and their related care limits the movement of female mammals relative to that of males (Greenwood, 1980). Intense focus on hunting and the nature of dominance in tribal groups may have biased a generation of anthropologists to seeing this phenomenon and understanding how it may lead to differences between phylogenetic patterns of genetic characters in males and females. The demographic consequences of sex differences on dispersal and the genetic characteristics of populations have been hinted at in Linda Vigilant's study of !Kung mtDNA hypervariable control region sequences (Vigilant et al., 1989).

Imagine that this geographic grid for existing humans could now be sunk in the ground and human skeletal samples from different shallow geological strata taken for the same distances. DNA extracted from these ancient donors would yield extensive information about philopatry, persistence of biological lineages, adaptation, and cultural norms determining kinship or inclusive group. It is possible that a very different and more satisfying level of interaction would exist between biological and cultural anthropologists if this data were available.

How ironic it is that our abilities to perform complex feats of molecular analyses have now outstripped our abilities to conduct human biological research in ways that are culturally sensitive to the concerns of aboriginal peoples. Remains of non-European populations, usually collected without their permission in the nineteenth century for comparative studies of growth, development, disease pathologies, and aging, now taunt physical anthropologists. These precious biological clues to human diversity and history represent much more than the loss of irreplaceable research materials when they are returned to descendants for reburial. They also represent the loss of hope for a better future. New, nondestructive methods of DNA extraction may encourage relatives of ancient donors to consider the information they will gain about their own past before they decide that nothing of significance, besides increased hatred, would occur if they too allow the bones of their ancestors to be used for scholarly studies.

6

Evolution

Thomas Jukes

The overriding theoretical perspective in biological anthropology is evolution. The final paper of this section by Tom Jukes (University of California, Berkeley) reviews the progress in refining evolutionary thought, incorporating the modern synthesis with molecular biology. One of Jukes's major contributions to this field, which he reviewed in his paper, "Silent Nucleotide Substitutions and the Molecular Evolutionary Clock" (Science, 1980), introduces the concept of the "neutral" mutation with its far-reaching consequences to human evolutionary studies. Jukes, in the chapter that follows, discusses how recent discoveries in molecular biology have impacted the study of evolution.

Evolution became a science during the nineteenth century. Jean Lamarck proposed in 1809 that evolution was a "drive towards perfection" and that it included the inheritance of acquired traits, which was shown to be incorrect. In 1858, Charles Darwin and Alfred Wallace proposed the theory of evolution through natural selection (Darwin, 1859; Brooks, 1984). Darwin implied that all living organisms might be traced back to a single ancestor from which they had branched, by variation, into all life forms.

A struggle for existence, according to Darwin, resulted in natural selection of species that had the best traits for surviving and reproducing. More recently, biologists developed the science of population genetics, in which a pool of variations of a species enables selection to take place.

Progress in biology led to the discovery of mutations, from which new types of organisms suddenly appear. Around the same time, Gregor Mendel discovered that hereditary factors behave as units transmitted by parents to offspring. Both findings broadened and substantiated the theory of evolution. Mutations provide for variation that can then be subjected to natural selection, and Mendelian genetics, which was unknown to Darwin, led to the discovery of genes, which are carried in the chromosomes of each living cell. Mutations act on genes to produce new strains of organisms. Many mutations produce harmful effects, and these lead to death or to elimination by natural selection. Today, through molecular biology, mutations can be studied in DNA molecules.

Darwin's proposal for the origin of species was supported by many of his observations, including those of finches and domestic dogs. The finches of the Galapagos Islands were of different species on the different islands. Darwin reasoned that they had changed over time from an original common ancestor. He also noted that dogs have been artificially selected into various breeds with enormous differences, such as those between a Great Dane and a toy poodle. The same is true of garden plants. If such change could appear in a short time under domestic conditions, Darwin postulated, then it should be possible for greater variation to occur over longer periods of time leading to new species.

The biggest bombshell in the history of evolutionary science exploded in 1953, although very few realized it at the time. The event was the discovery of the molecular structure of deoxyribonucleic acid (DNA) by James Watson and Francis Crick. They showed that heredity was a molecular process resulting from the sequence of four variables, nucleotides, which are abbreviated as A, C, G, and T in DNA molecules. The order in which the variables occur and the length of the DNA molecules, together, are responsible for inherited differences in all living species. If the sequence of human DNA were typed out as A, C, G, and T, in this letter size, it would be five thousand miles long. Changes in the sequences of DNA in various organisms, occurring through millions of years, produce evolution. The knowledge lacking was, first, what was the genetic code that translated the sequence into the proteins of which organisms are composed? And, second, how could the sequences be measured or "read" in the laboratory? Both of these problems have been solved. As a result, there is a flood of information on molecular evolution. Instead of speculating about our ancestry, we can now measure the number of differences between our proteins and those of yeast cells, thus mathematically showing, that yeast and humans share a common ancestor which lived about 1.2 billion years ago. Even our oncogenes, which are related to cancer, are found in similar sequences in yeast.

A family tree of life can be constructed by comparing the nucleotide sequences of ribonucleic acid (RNA) molecules in the small packages called ribosomes that are present in all living cells.

The tree shows us on a small twig shared by fish. Further along the branch are marine species such as starfish and sea urchins. Another large branch leads to the protozoa, such as *Giardia* and the parasites that cause sleeping sickness. Comparisons of ribonucleic acid sequences are based on the so-called molecular evolutionary clock. The clock depends on two observations: first, DNA replicates *almost* perfectly; second, the few imperfections that occur in its replication are a major driving force in evolution. These imperfections become "point" mutations, the term for substitutions of one base by another, such as T by C. These substitutions occur essentially at random and accumulate at an average rate of about five per billion nucleotide sites per year in DNA.

Family trees, also called phylogenetic trees, can be constructed by comparing amino acid sequences using the same protein that occurs in many species with slight to marked differences in its amino acid sequence. Each amino acid is coded by a sequence of three nucleotides which is called a *codon*. Most of the DNA strand consists of "non-coding" regions.

The DNA discovery showed that heredity takes place in all living organisms by reading the sequence of "words" in a four-letter language of nucleic acids. The hereditary information is maintained by duplicating this sequence and passing it to the next generation. Also, each generation receives the information from its parent, going back through time to the first molecules of DNA more than 3 billion years ago. If DNA ceased to exist, all life would become extinct, just as it does for a single species which fails to reproduce. Evolution can take place only through changes in DNA. Human beings are different from other forms of life only because of the number and order of the basis A, C, G, and T in their DNA molecules. The same rules apply for human heredity as for that of horses, daffodils, and bacteria. It is no wonder that in 1953, when the structure of DNA was discovered, Francis Crick, in the words of Jim Watson (1968), "winged into the Eagle at lunch to tell everyone within hearing distance that we had found the secret of life."

However, the work that remained was still a staggering task for biological scientists. The language of proteins contains twenty letters, the amino acids which combine to make them up. The four-letter language of the DNA, then, had to be translated into a twenty-letter language of the proteins. The four letters in DNA molecules had to be put in a sequential order, and this was achieved in 1977 by the English Nobel laureate Fred Sanger and by two Americans, Alan Maxam and Walter Gilbert.

In evolutionary studies a new science was initiated, that of molecular evolution. This science is based on the principle that molecules of DNA with similar sequences must have a common ancestor, because the laws of chance say that if two long sequences are sufficiently similar they cannot have arisen independently by accident. It studies evolution by comparing the DNA and protein molecules of many forms of life. Its achievements have fully confirmed the theories of Darwinian and neo-Darwinian evolution.

The study of evolution is a great intellectual adventure. The evolutionary process has been going on for more than 3 billion years, but we have been aware of it for only 150 years. Therefore, we have many unanswered questions about it. Nevertheless, scientists agree that evolution is the unifying principle of biology.

But what of human beings, with their languages, music, literature, ethics, religions, art, technology, superstitions, and knowledge? Most of these characteristics are the result of culture and education, not of heredity. It is a fact that every time a new organism starts its life, its inherited characteristics have passed in a thread through the eye of an ultramicroscopic needle. We now seek, through the Human Genome Project (mapping of the DNA code), to decipher our thread of life.

The Neutral Theory of Molecular Evolution

The neutral theory of molecular evolution says that nucleotide substitutions in DNA inherently take place as a result of point mutations followed by random genetic drift. If there are no selective constraints on these substitutions, the rate at which they are incorporated reaches a maximum value of about 5×10^{-9} substitutions per site per year. The rates are slower than this in genes that code for proteins, and these slower rates depend upon the rate at which the protein evolves.

The neutral theory was proposed during the 1960s when sequences of amino acids in proteins became known. These sequences, however, did not reveal the so-called silent changes which might have occurred at the third position in codons for the amino acids. Later, sequences of nucleic acids became available and provided further evidence for neutral changes.

In 1961, differences in the amino acid content of total protein in bacteria were found to be related to the base composition of the DNA in the bacteria. This was reported by Sueoka (1961) before the genetic code was known. He found that organisms such as *Bacillus cereus* whose DNA is high in A plus T are also higher in the amino acids isoleucine, lysine, phenylalanine, tyrosine, aspartic acid plus asparagine (Asx), and glutamic acid plus glutamine (Glx). In contrast, high G plus C organisms, such as *Micrococcus lysodeikticus*, are higher in alanine, arginine, glycine, and proline. In his 1961 article, Sueoka stated:

> When the GC content of two organisms differs appreciably, it is unlikely that a protein formed in one cell will be similar in primary structure to any found in the other. This also applies to enzymes of identical function with the exception that the active site may be similar but the dispensable parts of the molecule will be quite different.

When the code became known, it was found that isoleucine, lysine, phenylalanine, and tyrosine, but not Asx and Glx, had codons in which the first two positions consisted of A and T. Alanine, arginine, glycine, and proline had codons with C and G in the first two positions. The directional mutation studies by Sueoka revealed that many changes in codons could be produced without changing amino acid sequences. These would be neutral changes, and directional mutation pressure would favor such changes. An additional group of mutational changes would take place when amino acids were replaced in evolution without changing the functions of the proteins. Sueoka did not publish further on the subject until 1988.

Some of the first experiments with proteins indicated that neutral changes could exist. In the 1950s, Sanger and his colleagues discovered by studies with insulin that proteins were made of polypeptide chains with amino acids in a definite order (Jukes, 1966). They also found that insulins from different animals were different from each other because of amino acid replacements. For example, an alanine at one position in bovine insulin is replaced by threonine in human insulin. Nevertheless, since bovine insulin is used to treat human patients, amino acid substitution is neutral in its effect.

In 1966, I proposed that cytochrome *c* molecules in different species gave evidence of neutral changes (Jukes, 1966). Cytochrome *c* is a protein with about 104 amino acids and it is present in all eukaryotes. Its function is to transport electrons and, because of this, it contains an atom of iron. Cytochrome *c* in dogs differs from cytochrome *c* in horses in about 10 of the amino acids in its chain of 104. Yet the cytochrome *c* in dogs and horses has the same function. I suggested that it was possible that the two cytochrome *c*'s had been passively carried along as dogs and horses evolved separately from a common ancestor, in which case separation of the two species would be followed by changes in the two cytochrome *c* molecules. My conclusion was "the changes produced in protein by mutations would in some cases destroy their essential functions but in other cases the changes allow the protein to continue to serve its purpose."

The geneticist James Crow proposed an "infinite allele model" in evolution. This was described by Kimura and Crow in 1964 as follows:

> It is known that a single nucleotide substitution can have the most drastic consequences, but there are also mutations with very minute effects and there is the possibility that many are so small as to be undetectable. . . . We propose to examine some of the population consequences of such a system if it does exist.

The possibility of silent substitutions (neutral mutations) in genes became evident when the genetic code was deciphered. By 1965, even before knowledge of the code was complete, it was found that an amino acid could have a code or codes ending with G or C, or with A or U, as a nonspecific base that did not change its meaning. I pointed out in 1965 that an amino acid sequence could be coded by a series of codons with 73 percent G plus C, or the same sequence could be represented by a series of codons with a base composition of about 40 percent G plus C. The difference between the two sequences of codons was in the silent third position of the trinucleotides. Therefore, the directional mutation studies by Sueoka showed that many changes in codons can occur without changing the amino acid codon. These are neutral changes and directional mutation pressure favors such changes. An additional group of mutational changes can take place when amino acids are replaced, through the evolutionary process, without changing the function of a protein.

Jack Lester King, a population geneticist, and I started to collaborate in 1968, and we wrote a long manuscript which we entitled "Non-Darwinian Evolution." We challenged George Gaylord Simpson, who said in 1964, "It seems highly improbable that proteins, supposedly fully determined by genes, should change in a regular, but non-adaptive way." We said, in contrast, "natural selection is the editor, rather than the composer of the genetic message. One thing the editor does not do is remove changes that it is unable to perceive." We concluded that "the stream of spontaneous alterations in DNA, continuously fed into the genetic pool, should include far more acceptable changes that are neutral than changes that are adaptive." In addition to the "acceptable change," of course, would be a large number of deleterious mutations that would not be acceptable.

In 1968, King and I sent our manuscript to the magazine *Science*. It was rejected by both reviewers, but we appealed against the rejection and it was finally accepted (1969). Our article challenged the established idea that evolution took place solely through natural selection. We received much criticism, but in 1983, our publication was chosen as a "Citation Classic" by *Current Contents*.

Our article had been preceded by Kimura's article (1968), "Evolutionary Rate at the Molecular Level." He pointed out the existence of the molecular evolutionary clock in amino acid sequences of three proteins: hemoglobin, cytochrome *c*, and triosphosphate dehydrogenase. He then assumed that the entire genome evolved at the same rate as these examples. In contrast, King and I (1969) pointed out that "probably not much more than 1% of mammalian DNA codes for proteins." Even this conservative estimate is higher than today's figures of 50,000 to 100,000 genes in the mammalian genome. This would represent only about 0.03 percent of the genome.

Kimura then calculated the rate of base pair substitution within a mammalian genome as one every two years. This should be only about 0.2×10^{-9} per nucleotide per year, a figure far lower than today's estimates. He concluded that "most mutations produced by nucleotide replacement" must be almost neutral in natural selection, and that they occur "at the rate of roughly 0.5 per year per gamete." He assumed that "the entire genome could produce more than a million enzymes," which is far more than present estimates. He estimated a higher mutation rate in *Drosophila* than in humans.

Kimura, in 1988, described the details of how he came to propose the neutral theory in 1968. He disavows any collaboration with Crow. He says that Tomoko Ohta supplied him with some estimates of the rate of amino acid replacements in the actual course of evolution, and that, when he extrapolated these rates to the whole DNA of the mammalian genome, he "was surprised to note that it amounted to at least one base substitution every two years." He notes that at the time when he wrote his 1968 paper:

> I was accustomed to the "panselectionist," and emotionally I had difficulty in believing the preponderance of selectively neutral allele. Only scientific reasoning compelled me to assume selective neutrality of allele was involved.

He also remarks that he did not imagine that his paper would lead to "such heated controversies, and that my subsequent debates with anti-neutralists would give me such anxiety." It is evident that Kimura was unaware of my publications in 1965 and 1966 and that he arrived at the neutral theory by a different route.

The Molecular Evolutionary Clock

This term was introduced by Zuckerkandl in 1965, because he noted that sequences of amino acids in hemoglobin molecules accumulated differences at the same rate that the organisms containing the hemoglobin diverged from each other during evolution. The molecular evolutionary clock is therefore an adjunct of the neutral theory, which takes the clock a step further by pointing out that such amino acid replacements become fixed at a uniform rate by genetic drift. It is obvious that the divergences would not be uniform with respect to time unless the replacements were produced by a process that was independent of speciation. Max Perutz (1983) has commented that

> the structural evidence suggests that most of the amino acid replacements between species are neutral or nearly so, caused by random drift of selectively equivalent mutant genes, and that adaptive mechanisms generally operate by replacements in a few key positions.

This statement by Perutz is a neat summary of the neutral theory, and illustrates that the molecular evolutionary clock is a result of neutral changes in evolution.

Perspectives on Human Evolution

There are two viewpoints regarding evolution. The first is that everything was made in an unchanging form by an all-powerful Creator. This perspective was upset by the publication of Darwin's *Origin of Species*, although many still adhere to the former. The second viewpoint is that of Darwinian evolution. Recent discoveries have led to the study of evolution by comparing the sequences of proteins and nucleic acids in various organisms. These sequences differ in diverse species in a systematic way. I will discuss the effect molecular biology has had on Darwinian evolution.

One of the first publications on the sequential analysis of proteins (Tuppy, 1958) compared a small portion of the amino acid sequences of cytochrome *c* from vertebrates, insects, and yeast and showed that they were similar. Cytochrome *c* changes very slowly during evolution probably because it interacts with two other proteins, cytochrome *c* oxidase and cytochrome *c-1*, which tend to preserve cytochrome *c*'s three-dimensional structure.

Some molecules, such as those of riboflavin, nicotinamide dinucleotide and other coenzymes, are identical in all living organisms. In contrast, cytochrome *c* molecules, while quite similar, differ in various eukaryotes, and the differences increase with phylogenetic separation. What is the evolutionary explanation for the differences?

When the complete sequences of cytochromes c from yeast and human beings are aligned, about 56 out of 104 amino acids are identical in corresponding sites. The deduction from this is that yeast and human beings have a common ancestor. It is from such beginnings that the science of molecular evolution arose. Phylogeny could be measured arithmetically by comparing the sequences of cytochrome c or of hemoglobins. The phylogenetic trees constructed by these (see figure) are in accord with classical phylogeny as deduced from morphology and the fossil record. Molecular phylogeny, however, extends far back beyond classical phylogeny. Cytochrome c_{551} from the bacterial species *Rhodosporillum rubrum* has 41 percent of sequence identity with human cytochrome c.

Comparisons of other homologous proteins, such as hemoglobins, histones, insulins and fibrinopeptides, show that the rates of sequence divergence vary widely. Histone A molecules of cow and pea are identical, even though mammals and green plants have had separate ancestors for more than a billion years. The short fibrinopeptide A chains of cow and sheep differ by 26 percent in nineteen amino acid sites, and the hemoglobins of cows and horses differ by about 13 percent, while their cytochromes are only 3 percent different.

Neutral substitutions give rise to the molecular evolutionary clock. The "clock" is based on the premise that DNA replicates almost perfectly, but the few imperfections in its replication are a major driving force of evolution. These imperfections become point mutations which are substitutions of one base by another, such as T by C. These substitutions occur essentially at random and accumulate at a rate of about 5×10^{-9} per nucleotide site in the regions of DNA that have no effect on the sequence of amino acids in proteins. Such regions include non-coding regions, and also most of the third nucleotide sites on codons, which can change "silently" without affecting the code. For example, GGA is a codon for glycine. It can change silently to GGC, also a code for glycine, but a change from GGA to GCA is a change from a glycine to an alanine codon. This is a replacement change, not a silent change. The effect of replacement changes is to bombard proteins with amino acid changes. This also takes place at a steady rate, but at a rate lower than that of substitutions of nucleotides at non-coding sites. Many replacement changes are lethal, and so do not persist. Others are neutral or near-neutral so that the protein can continue functioning. Because the changes are randomly scattered along DNA molecules,

Evolution of Life

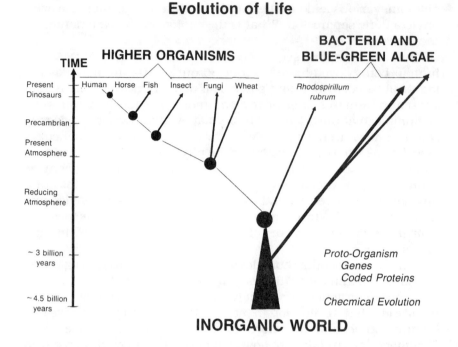

Evolutionary divergence as deduced from sequences of amino acids in molecules of cytochrome *c* in various organisms (from Dayhoff, 1972). The time scale is derived from radioactive dating of rocks. In molecular evolution, there are no "missing links" or undiscovered "intermediate forms." Changes in the sequences accumulate during time at an approximately steady rate, measured by the molecular evolutionary clock (see text).

they occur at different sites in different species. As soon as divergence from a common ancestor starts, this differentiation between the two species is initiated. The consequence is the molecular evolutionary clock, which is a real and measurable process of evolution. The rate of the clock is fairly constant for a given protein, especially when measured over long time intervals.

Evolution cannot proceed if molecular changes are only neutral or deleterious. There must be a few changes that are beneficial or adaptive. Evolution occurs through gene duplication: one of the duplicates retains its original function and the other duplicate acquires a new role, accompanied by nucleotide substitutions. Doolittle (1987) has commented that sequence data themselves generate new knowledge. This new knowledge is providing a remarkable picture not only of how living systems have evolved, but also of how they operate. Nowhere in the biological world is the

Darwinian notion of "descent with modification" more apparent than in the sequences of genes and gene products. We can see now that the predominate mode of gene evolution has been "duplication and modification."

Molecular biology shows that the process of evolution is continuous. Changes take place in regulatory genes as well as in protein-coding genes. Mitochondrial DNA changes more rapidly than nuclear DNA, because mitochondria do not contain the repair enzymes that correct nucleotide substitutions. Mitochondrial DNA, therefore, is a sensitive measuring index for studying human phylogeny and human evolution. Allan Wilson and his collaborators have used mitochondrial DNA for comparing ethnic and geographic relationships (see Cann, this volume).

Evolution is best appreciated by contemplating the facts of life. DNA must pass from generation to generation, or life will become extinct. Molecules of DNA, therefore, form a continuous link to the beginnings of DNA-based life, stretching back some three billion years. In contrast, living organisms, including ourselves, have a short existence, during which they are the means by which DNA is transmitted to the next generation. We measure history by centuries, recording the rise and fall of nations and civilizations. Evolution has a time scale that is beyond our experience or comprehension, the smallest unit being one million years. As the cliche satirically states, "Human beings are the DNA molecule's way of making another DNA molecule." We are at the mercy of DNA, which carries in its sequence the results of natural selection.

III

FOSSILS

*Every discovery of a fossil relic which appears to throw light
on connecting links in man's ancestry always has, and always
will, arouse controversy, and it is right that this should be so,
for it is very true that the sparks of controversy often illuminate
the way to the truth.*
—Sir Wilfred LeGros Clark from Huxley Memorial Lecture,
    Bones of Contention, November 26, 1958.

We have seen how the various theories about the human past
have been changed and been limited by the discovery of more and
more fossils. The "tide of discovery" has been very important in
suggesting that some views of human evolution are much more
probable than others.

One kind of understanding of the past comes from the fossils: the
record in the rocks. Just as was the case with questions concerning
time, continental drift, or comparisons of human beings with other
animals, so the early descriptions of fossils as remains of former
life were often ridiculed.

When a mammoth skeleton was discovered in Germany in 1696,
a school teacher named Ernst Tentzel compared the bones to those
of an Indian elephant and showed that the fossil and the bones of
the modern elephant were the same. One might have thought that
this was very convincing evidence, but Tentzel was tried by the
faculty and the fossil judged a freak of nature (Stirton, 1959:98–99).
Just as the sight of the Grand Canyon did not necessarily lead to
geological interpretations, the often-whole and well-preserved fossils
did not lead to any notion of the history of life. The climate of opinion
at the time when the first fossils were being discovered prevented
their recognition or acceptance as remains of former life.

Before 1960, there were very few fossil humans, and their dating
was uncertain. A common arrangement of the known fossils can
be seen in Sir Arthur Keith's Antiquity of Man (1915). With these
dates, which were thought to be correct, anatomically modern man
separated at the beginning of the Pleistocene, Neanderthals existed
in the Pliocene, and Javas belonged to the Miocene. Each discovery
was placed on the end of a branch and the times of separation were
deduced from the anatomy. Each new fossil was also placed in a
new genus. This kind of arrangement faded slowly with the
influence of new fossils, more accurate geological understanding,
and the synthetic theory of evolution. Now there are hundreds of
fossils, many accurately dated, and more are being found all the
time. It is no longer feasible in introductory textbooks to list the
individual specimens as was the custom until recently.

As of today, only a few fragmentary hominid fossils exist that are more than four million years old. In Africa, however, there exists a large fossil sample dating from three million years ago. In chapter 7, Noel Boaz describes this extensive sample, which spans from Ethiopia to South Africa. On the basis of numerous specimens taken from this sample, bipedal locomotion was well-established before three million years ago. The innominate (pelvis) bones from Hadar in Ethiopia and Sterkfontein in South Africa are close to identical and suggest that bipedal locomotion is the fundamental adaptation of humans. In this time period, there are skulls with very small cranial capacities (about 450 cc) and clear anatomical evidence of bipedal walking. A number of transitional forms link these early, gracile australopithecine fossils with the fully human, larger-brained, and taller *Homo erectus*.

Developing parallel to the gracile australopithecine forms is a more robust hominid creature. It survived at least a million years, was fully evolved by two and a half million years ago, and over-lapped *Homo erectus* for more than half a million years. By one and a half million years ago, *Homo erectus* is dominant, cranial capacity has doubled in size, and tools include bifacially flaked stones of the Acheulian tradition.

The direct fossil evidence for bipedalism has been debated for a long time. The root of the problem was the limitations of the fossil evidence. For example, when the first pelvis of *Australopithecus* was discovered in the caves at Sterkfontein, it was claimed by some as proof of bipedalism. Others stated that the pelvis was too modern to belong with the skull and that the specimen must have fallen in from a higher level. Today, after the discovery of many speci-mens, it is hard to believe that such controversies actually took place, but the beliefs about what should be found were so strong that they retarded the acceptance of the australopithecines as human (man-apes? early man?), as bipeds, as tool users, or as hunters ranging widely over Africa.

In retrospect, it is clear that the order of discovery exerted a great deal of influence on controversies which have lasted from Raymond Dart's initial description of Taung (the site of the first australopithecine find) in 1925 to the present day. I think we may see reality more clearly if we imagine the course of history as having been quite different. Suppose that a skeleton a little more complete than Johanson's Lucy had been the first find and that modern dating methods had been available. Then it would have been known that, at three million years, the "man-ape" creature had a very human pelvis, human thigh bones, and human feet. It would have been certain which parts went with what and that the creature was bipedal (albeit, a biped with a small brain and long arms). It is the finding of pelvis, knee, and foot bones together which confirms the diagnosis of bipedalism. Finding much of the skeleton of the same individual is that which gives confidence in an accurate reconstruction, and one relatively complete find is worth more than fifty years of often acrimonious debate.

Tools also allow the reconstruction of behavior, especially when assemblages are associated with accurately dated animal bones as J. Desmond Clark describes in chapter 8. Because the basic theory used in analysis is that of natural selection and because selection is for behavioral success, behavioral reconstructions are important and must be interpreted. Those populations whose behavior was the most adaptive left more descendants and thus determined the course of biological evolution.

Increased brain size follows long after evidence of tool making. Furthermore, Acheulian tools, which are difficult to make, follow simpler tools by at least a million years. It looks as if the successful life that the tools facilitated acted in a dynamic feedback relation with the evolution of the brain. What is seen in the human cortex mirrors evolutionary success. Just as the proportions of the human hand, with its large, heavily muscled thumb, reflects selection for success in tool using, so the brain reflects the success of hand skills. In the last hundred thousand years of human evolution the brain did not become any larger; however, technological advances increased geometrically.

Origins of Hominidae

Noel T. Boaz

*Any understanding of evolution must contain an appreciation
of the fossil and archaeological record. In the first chapter of
this section, Noel Boaz (Institute of Human Evolutionary
Research, George Washington University) brings the record of
the earliest fossil discoveries up-to-date. Having completed
extensive field research in Ethiopia, Libya, and Zaire, Boaz
addresses the recent paleoenvironmental factors which contributed to the evolution of humans. His recently edited
volume,* Evolution of Environments and Hominidae in the
African Rift Valley *(1990), and his review of the hominid fossil
record (*Yearbook of Physical Anthropology, 1988) are further,
important contributions to the field.*

Understanding the origin of the hominid family is one of the most
long-standing, and indeed almost classic, quests of modern science.
The questions surrounding humanity's emergence from the natural
world and its evolutionary divergence from its "animal" ancestors
were implicit in Charles Darwin's *Origin of Species*, first published
in 1859. Darwin's 1871 *Descent of Man* made many of the questions
explicit as regards extending the theory of natural selection to the
human species. As more refined research methodologies, both in
the laboratory and in the field, have been developed, research
questions have become progressively more focused. But the
specifics of the origin of the hominids remain as elusive as ever.
If there has been progress since Darwin's time, it has been in the
arena of reducing the possible hypotheses. We have yet to arrive
at an empirically defensible scenario for how, where, or when
hominids arose.

As paleoanthropologists have discovered successively older and
more primitive fossil remains of hominids, the definitions of the

family Hominidae, and its distinction from apes, have had to be constantly reassessed. Past definitions of *hominid* have been based primarily on characteristics of *Homo sapiens*, our own species, which is the best-known member of the family. Linnaeus' original (1758) definition was simply "*Homo nosce te ipsum*" ("Humanity, know yourself"), but this exhortation is now inadequate. The fossil record preserves several species of hominids, which differ in varying degrees from modern humans. The line between human and ape has had to be spelled out in much more detail.

Definition of Hominidae

The ancient Greeks defined members of the human species as "featherless bipeds." The term *biped* split people off from all four-footed animals, and *featherless* removed people from the largest category of bipedal animals, the birds. However, today such living animals as the kangaroo, gerbil, and gibbon (when on the ground), and such fossil animals as the *Tyrannosaurus*, pterodactyl (when not in flight), and the *leptictideum* (a small marsupial mammal capable of running and hopping, discovered at the German Paleocene site of Messel) would all fall into this category.

Sir Wilfrid E. Le Gros Clark (1964) provided what has been the most widely accepted definition of Hominidae. Hominids are relatively large-brained and bipedal members of the order Primates. They have orthognathous facial skeletons that protrude less than apes' faces. Hominids' canine teeth are reduced in size in relation to their other teeth, compared to apes whose teeth are more uniform in size. However, as more primitive hominid fossils have become known, some of these distinctions have become less absolute. For example, the earliest hominids now known show a ratio of brain size to body size only slightly higher than that of apes. Similarly, the upper canines of these earliest hominids are quite large and frequently produce a small wear facet on the side of the lower third premolar, similar to apes. Nevertheless, these early hominids were well-adapted bipeds, and bipedalism still serves as the most useful distinguishing characteristic of the family.

The Origin of Bipedality

Bipedality in the hominid fossil record is incontrovertibly established by 3.6 to 3.8 million years ago by the preserved track-ways at Laetoli (Leakey and Harris, 1987). The origin of upright posture and locomotion, however, is still obscure despite intensive interest in this subject since the time of Darwin. There is a dearth of relevant fossil hominoid material between 4.0 and 15.0 million years ago, and a total lack of relevant fossil documents of the *Pan* or *Gorilla* lineages (Hill and Ward, 1988).

Despite the fact that bipedalism has been considered a critically important character in the recognition of the earliest hominids, it may in fact be a plesiomorphic (original, primitive or underived) character shared with the common ancestor of the great apes and hominids, or even with the common ancestor of all modern apes and hominids. For now, the presence of morphology indicative of bipedalism in a fossil Neogene hominoid should be coupled with other traits diagnostic of hominid status, such as relatively large molars compared to body size (*megamyly*), in order to confirm the family diagnosis (Boaz, 1983).

The late-divergence hypothesis of hominids and great apes (Greenfield, 1980), prompted in no small part by molecular clock results indicating a date of divergence between 4.0 and 8.0 million years ago (Sarich and Wilson, 1967), has now largely been accepted (Andrews and Cronin, 1982; Pilbeam, 1984; Boaz, 1983; Simons, 1989). This recognition of the recency of human-chimp-gorilla ancestry was hailed as support for the knuckle-walking hypothesis, i.e., that the locomotor pattern of the ape-human ancestor was the same or similar to that of living chimpanzees and gorillas (Washburn and Moore, 1960). This model assumes that knuckle walking, present in two out of three of the species in this clade, is plesiomorphic for the African great ape-hominid clade. Yet, detailed comparative anatomical study of the modern human hand (Tuttle, 1969) and anatomical studies of australopithecine and early *Homo* hand bone elements (Susman and Creel, 1979; Marzke, 1983; Susman et al., 1984) have consistently failed to show any remnants of knuckle-walking anatomical modifications.

Further doubt has been cast on this hypothesis by the molecular findings that the *Pan-Gorilla*-hominid evolutionary divergence was not a trichotomy, as was first postulated on molecular grounds, but instead consisted of an initial branching off of the *Gorilla* lineage. Most molecular studies now support this cladistic arrangement (Holmquist et al., 1988; Caccone and Powell, 1989). The implications for the knuckle-walking hypothesis are profound. If *Pan* and *Gorilla* do not share a common lineage separate from hominids, then their shared locomotor pattern of knuckle walking is homoplastic. Other considerations support this idea.

Although the fossil record for *Pan* and *Gorilla* is largely nonexistent (only a single fragmentary canine of "Gorillinae" has been reported from the Ugandan Pliocene by Pickford et al., 1988), it is likely that the common ancestor for the hominid, chimp, and gorilla clade was small bodied. In fact, the common ape-hominid ancestor was probably small bodied. This deduction derives from the observation that early hominids were approximately only 1 m tall (Jungers, 1982; Schmid, 1983; McHenry, 1986), that *Sivapithecus*, now considered cladistically close to the *Pongo* lineage (Pilbeam, 1984), was also a small, macaque- to baboon-sized hominoid, and that the only surviving hominoids that have not adapted to life on the ground, the hylobatids, are similarly small sized (3 to 12 kg). Thus, *Pan*, *Gorilla*, and *Pongo* all independently evolved large body size, and as they did so, independently evolved a terrestrial quadrupedal stance—knuckle walking in the African apes and fist walking in *Pongo*. This falling forward onto the front limbs for support may have been made necessary by an allometrically positive increase in upper body size and weight when overall body weight increased. In this model then knuckle walking and fist walking are not plesiomorphic but apomorphic (derived from ancestral or primitive characteristics) for each of the great ape clades.

If knuckle walking was not primitive for the ape-hominid clade, what locomotor pattern was? One possibility is generalized quadrupedal climbing, a type of locomotor adaptation reconstructed for *Proconsul* (Walker and Pickford, 1983). However, this hypothesis fails to address the origin of the synapomorphies (derived or apomorphic characteristics shared by two or more different organisms) relating to verticality of the trunk seen in all modern hominoids, and lacking in *Proconsul*. Another, more likely suggestion is that the common ape-human ancestor was a vertical climber and hanger when in the trees and a biped (like the gibbon) when on the ground. Bipedalism, the same terrestrial locomotion utilized by hylobatids and early hominids, would then be the plesiomorphic condition for the great ape-hominid ancestor. Knowledge of post-cranial fossils of African hominoids in the middle Miocene to early Pliocene is necessary to resolve this issue.

Knowledge of the Earliest Hominids

Despite intensive searching for over a hundred years, fossils which document the common ancestor of the ape-human clade have eluded discovery. Fossils which bear evidence to the earliest stages in hominid evolution are fragmentary and consist of only a few teeth, jaw fragments, and limb bone fragments. These fossils come from Africa and date from 4.0 million years ago back to as early as 11.0 million years.

An isolated upper molar tooth from Ngorora, Kenya, is the earliest possible fossil evidence for hominids in Africa. It is dated to about 11.0 million years ago. The specimen is similar to a modern chimpanzee except that it is larger in size and likely has a thick enamel capping on the crown of the tooth. Both of these characteristics differentiate the tooth from modern African apes and are similarities with modern humans. However, because thick enamel is a primitive character for the great ape-hominid ancestor (Boaz, 1983), it is not definitive as to the hominid status of the Ngorora specimen. Without associated post-cranial bones it is impossible to determine the body size of the Ngorora hominoid and thus decide whether its molar teeth were relatively enlarged, as in hominids. A second hominoid specimen found at Ngorora is a premolar that Hill and Ward (1988) have suggested represents the last surviving *Proconsul*. The Ngorora molar, however, belonged to a different species of hominoid. It was also different from the gorilla-like Samburu Hills hominoid, at least 1.0 million years younger. It may represent the earliest hominid or it may be one of several still poorly known late Miocene East African hominoid species. Another isolated molar tooth, also from Kenya, comes from the site of Lukeino, dated to about 6.0 million years ago. In overall appearance it is chimpanzee-like but, for the same reasons as for the Ngorora specimen, little more conclusive can be decided regarding this specimen.

Some researchers accept that the earliest fossil evidence for Hominidae is the Lothagam mandible, from near Lake Baringo, Kenya, and dated to somewhat over 5.5 million years ago. This specimen consists of a right portion of a jaw with the first molar and the root of the last premolar preserved. The thickness of the mandible and the squared shape of the molar, as well as a number of morphological details, show that this specimen was similar to later hominids and significantly different from apes. A second jaw discovered at Tabarin, Kenya, and dating to about 5.0 million years ago, supports the position that hominids were present in East Africa by the end of the Miocene epoch.

Ironically then, the earliest fossil evidence for Hominidae, which is defined on the basis of bipedal locomotor capacity, consists of dental and mandibular remains. There is no clear indication of whether the Lothagam and Tabarin hominids were in fact bipeds. To ascertain this from the fossil record, one requires relevant portions of the post-cranial skeleton, preferably lower limb bones.

The earliest possible lower limb bone that may be hominid is the end of a fibula from the Libyan fossil locality of Sahabi, dated at about 5.0 million years ago (Boaz, 1987). The specimen, of dubious taxonomic affinities, shows the stout shaft and reduced fibular head seen also in humans. The first definitive hominid limb bone is a femur fragment from Maka at the northern Ethiopian site of Middle Awash, dated at less than 4.0 million years ago (White, 1984). It is clearly hominid in its long, straight neck and small greater trochanter. The straightness of the head, neck, and shaft indicates that weight was transferred in a more or less vertical manner, as expected in a biped.

Paleoclimate and Hominid Emergence

Africa in the early Miocene, 15.0 to 18.0 million years ago, was a forested place. Fossil sites on the edge of Lake Victoria, in eastern Uganda, and in eastern Zaire preserve abundant plant and animal fossils clearly indicative of forests and dense woodland (Kortlandt, 1972; Andrews and Van Couvering, 1975; Pickford, 1990). An abundance of small apes known as proconsulids, as diverse and plentiful as monkeys in a modern African or Amazonian forest, are preserved as fossils. By the middle Miocene some 5.0 million years later, apes had in general become larger in body size and less diverse in species numbers, and the rapid evolutionary radiation of the monkeys had begun.

These evolutionary changes are associated with vegetational and climatic changes. Middle to late Miocene faunas and floras from fossil sites not only in Africa but also in Eurasia, South America, and North America record a spread of grasses and woodland habitats. The apparent synchroneity of change has prompted the search for a common climatic cause. This has been considered to be the expansion of the Antarctic Ice Sheet. Isolation of Antarctica by continental drift and the beginning of an insulating circumpolar current caused greater refrigeration and build-up of ice.

To the north, tectonic movements at the Strait of Gibraltar made the strait shallower, allowing only the top layer of sea water, where dissolved salts are most concentrated, to flow into the Mediterranean Basin. Over several million years, the Mediterranean acted as a huge concentrator of salt, removing up to 6 percent of the salt from the oceans and becoming progressively saline itself in the process (Hsü, 1977). The decrease in global sea salinity affected Antarctica: as salinity decreased, the freezing temperature of sea water also decreased, thus leading to even greater formation of ice in Antarctica. By the latter part of the Miocene, about 6.1 million years ago, Antarctic ice had locked up water to such an extent that sea level had been lowered 40 to 70 m in the world's oceans. By about five and a half million years ago, the Mediterranean Sea was cut off at the shallow Strait of Gibraltar from inflowing Atlantic Ocean waters and began quickly to dry up.

The drying up of the Mediterranean is known as the "Messinian Event," after a locality in Italy. It was discovered by deep-sea drilling that uncovered extensive deposits of evaporitic salts, particularly gypsum and halite, left over from the hyper-saline pools occupying depressions in the desiccated Mediterranean Basin. At 5.0 million years ago, perhaps mediated again by tectonic movements at Gibraltar, the floodgates were opened to refill the Mediterranean bathtub in a waterfall of stupendous proportions.

Kenneth Hsü and colleagues first suggested in 1977 that the climatic changes accompanying the Messinian Event may have affected hominid emergence. They suggested that with the drying up of the Mediterranean, rainfall in the areas surrounding the basin would have been significantly lower and temperatures would have been cooler. These changes in turn would have encouraged the expansion of savannas and grasslands in Africa, habitats to which hominids became adapted.

Models of Hominid Emergence

Charles Brain (1981) has hypothesized that the Messinian Event and its attendant climatic changes not only *affected* but *effected*, hominid emergence. He points particularly to the time of divergence of the African ape and hominid lineages and to the timing of the Messinian Event and suggests that the congruence in time between the two argues for a causal association. Brain argues that the Messinian Event accounted for global cooling that occurred during the time period of 5.0 to 6.5 million years ago. He believes that due to decreased precipitation forests in Africa may have broken up at this time and become patchier, thus accounting for separation of ancestral hominid and ape populations.

Another school holds that regional tectonic and more localized climatic events were of greater importance in hominid divergence. Adrian Kortlandt (1972) and Martin Pickford (1990) have both suggested that the formation of the African Rift Valley system, particularly the Western Rift, were important determinants in splitting the ranges of ancestral hominids and apes. An increase in the height of the shoulders of the rift valleys prevented the dispersion of rain-bearing clouds into the inter-rift region from both the Indian Ocean to the east or the Zaire River basin to the west. The resulting arid rift floor formed a sharp demarcation to the West African forests over much of the extent of the Western Rift Valley. This vegetational change could have constituted a major disruption in population distributions and thus served as a potentially important evolutionary force.

Test Case 1: Sahabi

A site in northern Libya provided a test for the hypothesis of Mediterranean climatic and evolutionary change. The site was named Sahabi after a nearby fort in the northern Sahara Desert, now about 50 km inland from the Mediterranean coast. Fossils had been discovered at Sahabi through the 1930s and 1940s by Italian researchers led by Ardito Desio (1935) and Carlo Pettrochi (1952). The International Sahabi Research Project was formed in 1976, and fieldwork was carried out until 1981 (Boaz et al., 1987).

The geology at Sahabi revealed that a massive layer of gypsum underlay the fossil-bearing horizons. Geologists Jean de Heinzelin and Ali El-Arnauti (1987) interpreted this formation to represent the Messinian Event and, correspondingly, the overlying Sahabi Formation Beds to be of basal Pliocene age, about 5.0 million years old. Deposition had occurred in shallow tidal channels and alluvial deposits related to a large river. A deep channel near Sahabi had been detected by gravimetric research, extending to a depth under the desert of some 100 m. This river had cut down as the Mediterranean had evaporated and had flowed through a gorge the equal of the Nile. The river's source was a mystery.

Sahabi yielded abundant fossils of animals and plants. It has provided the best view that we currently have of the North African early Pliocene environment. It is one of the best test cases for Hsü's and Brain's models. One would predict that the fauna from Sahabi would differ appreciably from earlier Miocene faunas because the climatic changes that accompanied the Messinian Event would be expected to have been of significant effect in causing extinctions and promoting the evolution of new species adapting to the changed conditions. The evolutionist Elisabeth Vrba (1980) has argued that rapid evolutionary turnovers in species should be expected following major climatic change.

Contrary to expectations, the Sahabi fauna did not show a great degree of difference from preceding Miocene faunas. In fact, it was quite conservative. The carnivores looked very similar to those from the Miocene of Turkey, just across the Mediterranean. One of the most common large animals, a hippolike ungulate known as an anthracothere, was a holdover from the Miocene and had disappeared elsewhere in Africa by this time. The elephants and most other fauna resembled in many respects the pre-Messinian faunas from North Africa. There had not been the "punctuated" change hypothesized by some theorists.

Did this result indicate that the Messinian model for hominid divergence was incorrect? Not necessarily. Vrba (1985) introduced the concepts of *refugial areas* and *biotidal areas* in the fossil record.

Sahabi could have been one of the former: a locality insulated from climatic change by its large perennial river. One would see the effects of climatic change only in the peripheral species preserved by chance in the fossil record. The small mammals from Sahabi did indicate substantial aridity away from the river and the fossil wood showed fire scarring, as is seen in trees surviving savanna bush fires. Biotidal areas, on the other hand, are recognized on the basis of strong faunal responses to environmental change.

Further investigation of Sahabi is important to resolve a number of important questions that at present can only be answered there. If a fuller idea of the fauna were obtained, would it be possible to uphold the refugium model of Vrba? Did the large river at Sahabi ameliorate climatic change, and if so, where was its source? Some indications pointed to the headwaters of the Niger in far western Africa. What were the primates at Sahabi? We discovered the first fragmentary primate specimens from Sahabi but their identities cannot be established with any degree of confidence until more complete remains are uncovered. Work at Sahabi will resume as soon as political conditions permit.

Test Case 2: The Semliki Valley

The hypothesis that the African Rift Valley system has been responsible for the evolutionary divergence of apes and hominids has been one of the driving forces behind the Semliki Research Expedition to eastern Zaire, an ongoing international research project since 1982. The Semliki River runs between Lakes Rutanzige (formerly Edward or Amin) and Mobutu (formerly Albert) along the floor of the Western Rift Valley, eroding fossil-bearing sediments as it goes. Jean de Heinzelin's expeditions had investigated the Semliki Valley fossil deposits during the 1950s up until 1960 (de Heinzelin, 1955; Gautier, 1965).

Biogeography is one of the strongest arguments in favor of what we may term the Rift Valley vicariant hypothesis. (*Vicariant* refers to a population being split into two, resulting in the formation of two new species.) African apes today live west of (or on the margins of) the West African Rift, and all the known Plio-Pleistocene early hominid sites occur to the east of (or in) the West African Rift. At none of these fossil hominid sites are fossil apes found. This fact argues that apes in the past, as today, were limited to the lowland or montane forests of central and West Africa. They did not venture into the savannas and grasslands of East Africa. How did this ape-human split occur?

Some scholars have suggested that the uplift of the shoulders of the rift walls was sufficient to cause a rain shadow from both west and east, thus blocking rain from reaching the rift floor. This would have created a situation in which forests in the rain shadow area would shrink. Ancestral hominid populations might have be forced to cross open patches of land to get to other patches of forest. Through natural selection the first hominids would have emerged when they adapted entirely to the open patches and moved eastward, walking bipedally, into the expanding savannas. Ancestors of the chimpanzees and gorillas on the other hand would have retreated more deeply into the forests west of the Western Rift.

One of the goals of the expedition was to find out when the Western Rift had become a major forest-restricting ecological barrier. Thus, we could begin to understand whether or how this phenomenon might have had an effect on the ape-human split. Work began in the southern or Upper Semliki Valley which preserves sediments of Plio-Pleistocene age. Very few fossil vertebrate animals were known when the expedition started working in 1982. Today some fifty-one vertebrate species are known from the oldest levels, dating to about 2.0 to 2.3 million years ago based on faunal comparisons with radiometrically dated sites in East Africa. Up to now we have found no volcanic levels in the Western Rift that can be dated by the potassium-argon or argon-argon methods.

Analysis of the Semliki fauna shows a preponderance of savanna or woodland-adapted species, very similar to eastern Africa at this time period. No apes have been found, but we know that hominids were present because they left abundant stone tools, studied by archaeologists J. W. K. Harris and Alison Brooks (Harris et al., 1987; Brooks and Smith, 1987; Boaz et al., 1990). Studies of the fossil wood showed that many of the trees were also savanna-adapted species, such as *Acacia*, and some even showed the fire-scarring characteristic of savanna bush fires. However, the forests were not far away from the Semliki fossil sites in the rift valley floor, because rare pieces of fossil wood showed up that belonged to large forest species characteristic of central and West Africa. The invertebrate fauna studied by Peter G. Williamson (1985) confirmed that a major period of desiccation occurred in the Western Rift at this time.

The research in the Upper Semliki succeeded in demonstrating that the Western Rift had already become a major ecological barrier to forests and forest-adapted species by the Late Pliocene, about 2.0 million years ago. It became clear that to find the forest biomes in the Western Rift, from which earliest human ancestors may have emerged, we would have to look in older sediments. Deposits from the Lower Semliki, near Lake Mobutu, may hold the answer. They are of Miocene and Pliocene age and have yielded some fossils, unfortunately not diagnostic enough to ascertain the paleoecology or definite age of the deposits.

Choosing between Alternative Models

Which way of looking at the paleoecology of the ape-human split is more likely correct? Were global climatic cooling and increased aridity at the Messinian Event primarily the factors responsible for isolating incipient hominid ancestors, or was there a much more localized series of tectonic events along the Western Rift in central Africa that set the factors into play resulting in the divergence of the hominid lineage?

The results from Sahabi, the Semliki Valley, and other sites are far from adequate to answer this question. We will need much better resolution of the age of the various deposits, especially in the Western Rift, a broader view of the contexts of the fossil sites, both in terms of space and time, and much more data on the flora and fauna. Included within the latter category of course are the fossils of the apes, hominids, and their ancestors who are the central players in our models.

Any simplistic either-or answer is unlikely, however. A combination of global and regional, climatic and tectonic factors probably initiated the geographic and consequently the genetic isolation of proto-ape and proto-hominid ancestral populations. Although in modern parlance this splitting of an ancestral population to cause the formation of two new species is known as a *vicariant event*, it may in fact have occurred through a series of "events." A fluctuating curve of environmental change resulted in successively different rainfall and temperature patterns playing out over a changing landscape in central Africa. The shifting mosaic of forest and savanna moved and re-moved the animal and plant populations like chess pieces over a giant Rift Valley playing board. There was extinction of those populations that became stranded in a shrinking island of forest or savanna. And there was evolutionary change as natural selection worked to adapt surviving populations to new habitats, such as the plains for emerging hominids. The apes retreated to their ancestral habitat, the forest, even though future research may show that they underwent evolutionary changes as well.

The evolutionary split between humans and apes is one of the most venerable problems in science. However, by employing a multidisciplinary research approach that includes earth, biological and anthropological sciences, real headway is being made. Data are being gathered that are critically important in setting the paleoecological stage, in determining the temporal and geological frameworks, and in understanding the genetic and population changes that have occurred.

Cladistic Relationships

Early on there was some debate concerning the relationships of humanity with the African or Asian great apes. Ernst Haeckel (1906) considered the orangutan to be the most closely related ape to humans, while Thomas Henry Huxley (1863) considered the African apes, the chimpanzee and gorilla, to be the most closely related. Molecular data conclusively decided this issue in favor of the African apes. Fossil discoveries in Asia have also shown that orangutan-like morphology in the face and dentition had appeared by 8.0 to 12.0 million years ago.

The cladistic relationships within the hominid-African ape clade are much more difficult to resolve, however. Most early studies concluded that there was an unresolvable three-way split, a trichotomy, among hominid, chimp, and gorilla lineages, dating to between 4.0 and 8.0 million years ago (Sarich and Wilson, 1967). More recent evidence from DNA-DNA hybridization and mitochondrial DNA studies have consistently indicated an earlier cladogenesis of *Gorilla* from human-*Pan*. If true, this indicates that the traits long considered to be most diagnostic for the African apes, particularly knuckle-walking morphology and characters of the dentition, especially thin molar enamel, are in fact homoplastic, or developed in parallel. This issue will only be resolved by future fossil discoveries in this important time period of earliest hominid emergence.

The Australopithecines

Australopithecines (meaning "southern apes") are the first well-known hominids. The genus *Homo* evolved from early australopithecines. These hominids have sometimes been referred to as "ape-men" or "man-apes," allusions to their transitional position between humans and apes, but the later "robust" australopithecines are specialized species that became extinct without issue. Australopithecines are characterized as a group, and set apart from humans, by their small cranial capacity, a heavily built and usually protruding face, a large overall dentition, different pelvic structure, and strong grasping ability in their hands and feet. In contrast to apes, the australopithecines were bipedal, megamylic (large-molared), possessed of a smaller and functionally different canine tooth, and showed many anatomical details that ally them with more advanced hominids.

Controversy has surrounded the original australopithecine find, a skull of a juvenile hominid from a cave site quarried for lime in northern South Africa known as Taung, meaning "place of the lion" in the Tswana language. Nevertheless, the discovery is a singularly important event in the history of human evolutionary studies because it brought to light a totally unknown, though not unexpected, primeval state of human existence.

Dart (1925) suggested that Taung represented "an extinct race of apes intermediate between living anthropoids and man," a "missing link" to use T. H. Huxley's now-famous term. Dart named it a new genus and species, *Australopithecus africanus* and claimed a human ancestry for it. On the basis of the other fossil animals found at the site an early Pleistocene or late Pliocene age was postulated.

Only recently have doubts been cast on this dating of the Taung site. Geological extrapolations by Tim Partridge (1986) based on the rates that streams erosively cut valleys have been suggested to date Taung to only about 500,000 years ago, in the middle Pleistocene. However, renewed studies of the fauna associated with the site, including the monkey fossils (Delson, 1988), indicate a date of at least 1.0 million years ago. Other faunal indicators may place the age closer to 2.0 million years ago, an estimate that renewed excavation now underway by Phillip Tobias in the Taung breccia will test.

After Dart's original 1924 find, and the inevitable taxonomic problems associated with a juvenile-type specimen, there was much interest in recovering more australopithecine remains, this time of adults. A number of anatomists had pointed out that juveniles of many higher primates can closely resemble one another, while the adult forms are quite divergent. Robert Broom (1950) managed to find four bone-bearing cave sites, in addition to Taung, which did indeed yield remains of adult australopithecines, including a complete skull from a site known as Sterkfontein. These sites have continued to yield fossil bones to the present day. They date to between about 3.0 and 1.0 million years ago.

In East Africa, Louis Leakey (1951) started fieldwork at Olduvai Gorge, Tanzania (then Tanganyika), in 1931. Until recently it has been thought that the Leakey teams were unsuccessful in discovering early hominids until 1959, twenty-eight years later, when Mary Leakey found the skull of *Australopithecus boisei* (Louis Leakey, 1959). In fact, a lower canine tooth that Louis Leakey found in 1935 at the site of Laetoli, just to the south of Olduvai, sat in a museum tray in London until 1979, when it was recognized as the first specimen of an australopithecine discovered in East Africa (Mary Leakey, 1978; White, 1981).

Two of the most important eastern African sites for early australopithecines (*Australopithecus afarensis*) are those at Hadar, in the Afar Triangle, northern Ethiopia, and Laetoli, northern Tanzania. Hadar was explored by geologists Maurice Taieb and Jon Kalb in the late 1960s (Taieb et al., 1972), and fossil collecting and excavations were started there in 1973 by teams directed by Donald Johanson (1974). Laetoli had been known through Louis Leakey's early collections and the discovery by a German team under the direction of Ludwig Kohl-Larsen (1943) of a hominid upper jaw and single tooth in 1939. However, until the locality was intensively collected by a team directed by Mary Leakey between 1974 and 1979, its importance was not recognized (Mary Leakey, 1984). Twenty-four australopithecine fossils, as well as two trails of hominid footprints, are now known from Laetoli. Together, Laetoli and Hadar have given paleoanthropologists their best glimpse of the early australopithecines in eastern Africa, at a somewhat earlier time period than that of the South African sites of Sterkfontein, Makapansgat, and probably Taung.

Australopithecine Diversity

In his original paper, Dart proposed that a new zoological family, Homosimiadae, be created for its single contained species, *Australopithecus africanus* (1925). It was later pointed out that, according to the International Rules of Zoological Nomenclature, a family name must derive from a generic name of one of the species within it. After landmark anatomical studies by William King Gregory (1949) and Wilfrid E. Le Gros Clark (1947) in the 1940s, the Taung specimen and other South African australopithecines were generally considered to be bona fide members of the family Hominidae. These South African near-humans, however, were different and warranted some distinction from the genus *Homo*. It was proposed that they be placed in their own subfamily, the Australopithecinae, to be distinguished from the more advanced members of the family, placed in the subfamily Homininae.

Many paleoanthropologists consider there to be two types of australopithecines: a "gracile" species with a more lightly built skull termed *Australopithecus africanus* (sometimes *A. afarensis* is included in this category), and a "robust" species with a more heavily-built skull, known as *Australopithecus* (or *Paranthropus*) *robustus*, or in eastern Africa, *A.* (or *P.*) *boisei*. The gracile forms, by most reckonings, occur earlier in time than the robust and are considered by most paleoanthropologists to be ancestral, in a broad sense, to the genus *Homo*. Others have opined that the terms *gracile* and *robust* have outlived their usefulness, as even some early australopithecines, as at Hadar, may show quite "robust" cranial features.

Both African groups of early australopithecines are similar. They both possess relatively large canines, a tendency for non-bicuspid premolars (like apes), a high degree of sexual dimorphism and a face that is "dished" or depressed in the area around the nasal opening. Some differences have been pointed out between the eastern African sample from Laetoli and Hadar, on the one hand, and the South African hominids on the other. Most anthropologists have considered these differences to be of a magnitude to warrant a separate species name, *Australopithecus afarensis*, while others are of the opinion that the differences that are seen are to be expected in populations composed of individuals possessing their own somewhat variable characteristics.

Australopithecus africanus has been considered for some years to represent the hominid ancestral to the genus *Homo*. However, in their renaming of the Laetoli and Hadar australopithecines, D. C. Johanson and T. D. White (1979) suggested that only *Australopithecus afarensis* was ancestral to *Homo*, and that *A. africanus* was ancestral to the later robust australopithecines. The morphological similarity of the specimens in the two samples attributed to *A. africanus* and *Homo* seems to argue against this proposal, as does the consequent lack of an ancestor to the *Homo* lineage after 2.8 million years B.P. (before present), the upper date at the Hadar site. The earliest members of the genus *Homo* show a number of similarities to gracile australopithecines, and these similarities also argue for an evolutionary sequence from *Australopithecus africanus* to *Homo*.

There are no certain occurrences of australopithecines outside Africa. G. H. R. von Koenigswald (1955) reported some fossils (single teeth) from mainland China which he bought in Hong Kong and which he considered to be australopithecine. These specimens, however, are undated and are insufficient evidence to confirm the presence of the genus *Australopithecus* in China. Three mandibles collected in Java were considered by J. T. Robinson, an expert on the South African australopithecines, to be robust australopithecine (1953). However, most other anthropologists consider these specimens to be the remains of large individuals of early *Homo*.

The Appearance of *Australopithecus* in the Fossil Record

Australopithecus is known from both East and South African findings. The first major occurrence of gracile australopithecines is at Laetoli, in northern Tanzania. The site is some 50 km (30 miles) to the south of Olduvai Gorge and, as was mentioned above, was the site of the discovery of an australopithecine canine by Louis Leakey in 1935. Following the discovery of some fragmentary hominid fossils by Phillip Leakey, Mary Leakey reinstituted collection there in 1974 (Mary Leakey, 1978). A sample layer of volcanic ash from Laetoli was dated by Curtis to 3.7 million years ago, by the potassium-argon method (1975).

The hominid fossils from Laetoli were originally ascribed to the genus *Homo* by Mary Leakey (Mary Leakey et al., 1978). As such they represented the earliest known occurrence of this genus in the fossil record, fully 1.5 million years earlier than well-dated *Homo* fossils from other sites, such as Omo. The detailed description of the fossils by Tim White (White et al., 1981), however, revealed that they possessed a number of primitive traits, not expected in specimens belonging to the genus *Homo*. In traits in which the Laetoli fossils differed from *Homo*, they bore similarities to the South African early hominid fossils ascribed to *Australopithecus africanus*. White chose to emphasize certain apparent differences in the Laetoli specimens and, with Donald Johanson, ascribed the specimens to a new species, *Australopithecus afarensis*, based also on the combined fossil hominid sample from Laetoli and the Afar site in northern Ethiopia (Johanson and White, 1979).

In 1976, the remarkable discovery was made that one of the tuffs at Laetoli had preserved footprints of animals living when the ash was deposited, 3.7 million years ago. Among these animal tracks were trails of two hominids, walking side-by-side. These footprints of walking hominids constitute the earliest evidence of bipedalism in the hominid fossil record. In addition, White calculated that on the basis of foot size, the hominids were between 119 and 139 cm (3.9 and 4.6 feet) tall (1980). Charteris, Wall and Nottrodt in an ingenious paper (1981) demonstrated that by the estimated size of the hominids and the length of their stride, they must have been walking quite slowly, "strolling," across the African savanna. Mary Leakey (1979) suggested that by the difference in their size, the footprints were probably those of a male and a female.

The most abundant fossil evidence for early australopithecines comes from Hadar, located nearly one thousand miles (1600 km) to the north of Laetoli, in the Afar Depression of northern Ethiopia. The first hominid fossils to be discovered at Hadar were a distal thigh bone (femur) and the proximal shin bone (tibia), which fit together to form a partial knee joint (lacking only the patella). Donald Johanson (Johanson and White, 1979) showed that the angle at which the femur articulated with the tibia in the fossil from Hadar was like a hominid and unlike an ape. The femora in hominids slant downward to the knees that are quite close together, whereas the femora are aligned with the long axis of the tibia and the knees are widespread in apes. This was a clear indication from anatomy that the early australopithecines at Hadar were bipedal, a confirmation of the evidence from Laetoli.

Further discoveries at Hadar confirmed it as one of the most productive of hominid fossils sites. Among the sample, consisting of between 35 and 65 individuals, were the largely complete skeleton from Hadar Locality 288, nicknamed "Lucy," and a concentration of remains of some 13 individuals from Hadar Locality 333, nicknamed the "First Family." These remains constitute the most complete evidence that anthropologists have so far had of this most distant time period of hominid evolution. Potassium-argon dates have bracketed the age of the Hadar hominids between 2.8 and 3.6 million years old. Study of the fossils has revealed that early australopithecines possessed a high degree of sexual dimorphism, although some workers consider there to be more than one species at Hadar. However, australopithecine males, although larger than females, did not have proportionately larger canines, an important difference from sexually dimorphic non-human primates. This difference implies that canines had ceased to be used in intraspecific display and in male-male aggressive competition.

The Hadar hominid finds also confirmed estimates from the footprint evidence at Laetoli that the early australopithecines had been small. Lucy, for example, was estimated to have been about 1 meter (less than 4 feet) tall, although males would have been taller and larger.

The South African australopithecine discoveries have been somewhat eclipsed by the dramatic findings in eastern Africa. But as the excitement has worn off, anthropologists have begun to carefully compare the East and South African samples. Dating at the South African cave sites has never been clear since they have been based on relative ages as ascertained by the evolutionary stages of the fossil vertebrates contained in the assemblages. Nevertheless, the ages of Sterkfontein and Makapansgat can be confidently estimated to be between at least 2.5 and 3.0 million years old. Makapansgat appears to be somewhat earlier than Sterkfontein.

The temporal range of australopithecines, as determined by absolute and relative methods of dating, are 3.8 to ca. 2.5 million years for *Australopithecus afarensis*, ca. 3.0 to ca. 2.0 million years for *A. africanus*, and 2.5 to ca. 1.0 million years for robust australopithecines.

Australopithecine Morphology and Behavior

Although interest has been concentrated on hominid brain evolution since the first discoveries of fossil humans, studies of fossil endocasts (casts of the inside of the cranial cavity) have yielded disappointing results. The external form of the brain is very variable, making its study based on crania difficult. Additionally, too little is known regarding the function of the modern human brain for anthropologists to be confident in assessing the functional capabilities of early hominid brains.

The cranial capacity of *Australopithecus africanus* is known to lie between approximately 400 and 600 cc, roughly equivalent to the brain sizes of modern apes. There is one important difference, however. Body size in gracile *Australopithecus* is much smaller than in the modern gorilla or even chimpanzee. *Australopithecus* had a brain-to-body-size ratio that was larger than in modern African apes. A relatively larger brain implies that reorganization of neurons had taken place and that australopithecine behavior had become in some respects more complex and elaborate than that which we see in living apes.

Historically, brain size has figured prominently in the classification of fossil hominids. At one time it was proposed that a brain volume of 800 cc was a hominid "cerebral Rubicon," named for the Italian river crossed by Caesar in his conquest of Gaul and used figuratively as a "decisive step." Australopithecines, it was argued, fell below that value and thus should not be considered hominids. This argument not only failed to recognize the smaller body size of australopithecines (and thus relatively larger brain size) but was inflexible in disregarding the known principle that biological populations show variability in any trait. Modern humans, for example, show a range in brain capacity from under 1,000 to over 2,000 cc, and it is certain that past hominid populations showed degrees of variability.

Initially, there had been high hopes of understanding australopithecine behavioral capabilities from brain morphology as determined from cranial endocasts. Such attributes as speech, tool use, and sociality were expected to have observable neurological correlations. Unfortunately, although cells that perform certain functions are localized in the brain, the morphological and behavioral correlates could not be satisfactorily determined. Part of the problem lies in the variability of brain structures. Additionally, endocasts do not preserve precise surface detail of the brain, because they are actually impressions of the inside of the cranium, not the outside of the brain. While alive, the tissues which lie between the skull wall and the brain to cushion it (the meninges) obscure the smaller convolutions and surface features. In comparison to living ape brain morphology, the known gracile australopithecine endocasts may suggest a greater degree of folding and a larger number of convolutions. These features seem to predict greater neurological complexity in australopithecines.

Skull morphology is the most important criteria for recognizing *Australopithecus africanus*, because most of the important anatomical and behavioral adaptations of the species are reflected in the skull. Three general adaptations account for cranial form: brain size, erect posture (bipedalism), and use of the teeth.

Its relatively large brain gives *A. africanus* a somewhat globular head shape compared to modern apes. This shape is emphasized by the lack of the heavy ridges for muscular attachments seen in apes. One of the muscles of mastication, *m. temporalis*, is noticeably less developed in *A. africanus*. This muscle attaches to a slight ridge halfway up the side of the cranial vault. In gorillas and male chimpanzees the *temporalis* muscles of both sides of the head meet in the midline at the top of the skull, and a heavy ridge of bone, known as the sagittal crest, is formed.

The face of *A. africanus* is much more lightly constructed than in contemporary apes and the former shows a peculiar morphology—the nose region is depressed relative to the rest of the face. The functional significance of this "dished-face" morphology is not yet understood, although a likely explanation is that erect posture placed a premium on a depressed midfacial region for head balance, and the bone anchoring the enlarged molar teeth at the sides of the face increased in thickness. D. Pilbeam (1972) has applied the term *megadont*, meaning "large-toothed," to refer to the relatively large molar size of early hominids as compared to apes.

Overall, the gracile australopithecine face is orthognathous, or pushed in under the braincase, in a form similar to that of later hominids. The facial skeleton has moved backward and downward and the back of the braincase has rotated forward, a process known as *basi-cranial flexion*. Thus, the opening through which the spinal cord enters the brain, the *foramen magnum* ("great window"), is located halfway between the front and the back of the skull so that the head is balanced on the vertebral column. In the knuckle-walking apes the foramen is positioned more posteriorly and its opening is slanted more towards the back of the skull then in hominids—characteristics related to the more horizontal posture of these animals. Heavy neck muscles hold up and move the head. In *A. africanus* a reduced face, basi-cranial flexion, a centrally placed *foramen magnum*, and a lack of heavy neck musculature are features which reflect bipedality.

Teeth are the most abundant remains of the gracile australopithecines. Their structure parallels that of later hominids of the genus *Homo*: generally large and wide incisors relative to canines, canines functionally similar to incisors, and lower third premolars which do not wear or hone against the back of the upper canines, as in apes. In certain aspects, however, australopithecines are more primitive. The canines are generally larger than in *Homo*, the lower third premolar is less "squared" in appearance, the molars are generally longer relative to their width and they possess a complicated, wrinkled ("crenelated") surface pattern. Australopithecines, as all hominids, possess thick enamel on the top surfaces of the molars, a characteristic also shared with some apes, such as the modern orangutan. Thick enamel increases the functional lifetime of the tooth, especially when the diet is abrasive. The larger and crenelated molars of australopithecines provided them with increased surface area for grinding food, which probably had a lower protein content than that of later *Homo*.

The form and shape of the mandible in *A. africanus* has been of interest because in certain individual specimens it is markedly V-shaped. Interest has centered on whether or not this characteristic separates all *A. africanus* from later (or perhaps contemporaneous) species of *Homo*, bearing a more parabolic shape to their mandibular outline. There seems to be an intergradation between V- and parabolic-shapes in modern human mandibles and in early hominid mandibles found in the same locality. The most likely interpretation seems to be that *A. africanus* populations had individuals with V-shaped mandibles, recalling a more primitive condition, and parabolic-shaped mandibles, foreshadowing the condition generally seen in *Homo*. The shape of the mandible is not only a function of the width of the symphysis (the region in the front holding the incisors) but the distance between the condyles articulating with the skull base. Both of these characteristics are variable and unrelated to basic mandibular function.

The pelvic, shoulder-girdle, and limb-bone remains that have now come to light confirm that the gracile australopithecines were bipeds who did not use their upper limbs in locomotion. This adaptation seems to have differed somewhat from that of modern man. *A. africanus* had a relatively wider distance from the hip joint to the muscle attachments at the top of the thigh (femur) and a wider flare of the upper crest of the ilium. This arrangement provided a wide base of support for the lower limbs in a bipedal stance, as well as strong leverage in lifting the lower limbs during walking. In *Homo* the hip joint has moved closer to the top of the femur and lateral edge of the ilium, in order to expand the birth canal for larger-brained infants. The australopithecine morphology is now recognized as an efficient, albeit different, bipedal adaptation.

Australopithecine Paleoecology and Behavior

In his original paper, Dart pointed out the arid nature of the habitat in which *A. africanus* must have lived at Taung (1925). This environment would not have been inhabitable for an ape, assuming extinct apes had similar ecological adaptations to those of forest-living gorilla or chimpanzee of today. Further research has shown that *A. africanus* lived in both arid and more well-watered environments in eastern and southern Africa. It is clear that even without the use of stone tools and fire, gracile australopithecines were able to adapt to a variety of environmental conditions, but apparently only in Africa.

Within the past decade interest in hominid paleoecology has increased—partly because of the realization that environment is an important variable in determining social behavior in primates and partly because of the so-called single-species hypothesis. This hypothesis held that all early hominids (*Australopithecus* and early *Homo*) lived in a cultural-ecological niche with widely overlapping adaptive capabilities. Only one species could exist at any one time in the same area. If two species had been present, the principle of competitive exclusion would have come into play. By this principle one species would have driven the other to extinction by expropriation of environmental resources, the primary one of which is, of course, food. Discoveries since the hypothesis was proposed have now demonstrated that while there was only one species of gracile australopithecine living at any one time, later on there were certainly two species of hominids living contemporaneously (early *Homo* and *Australopithecus robustus*). Thus, the assumptions of the hypothesis, particularly those concerning the cultural capabilities of early hominids at this time, were incorrect.

The paleoecology of *Australopithecus africanus* is reflected in the types of animals, plants and geological conditions that are found associated with this hominid in fossil deposits. It is important to determine how these remains came to be buried together in order to understand how they might have been associated in life. The discipline that deals with burial and transport processes of fossils, taphonomy, has become an important part of the study of hominid paleoecology. Taphonomic studies have shown that hominids in the South African cave sites were probably the remains of carnivore kills, while East African open-air sites contain hominids that had died under a variety of conditions.

In the early South African cave sites, A. *africanus* has been found in association with a high percentage of bush-adapted, as opposed to grassland-adapted, antelopes, an association which indicates a more bush-covered habitat than the same area today. This paleoecological picture in South Africa is supported by similar vegetational reconstructions for two early australopithecine sites in Ethiopia (Omo and Hadar). The fauna from Laetoli, on the other hand, indicates markedly dry grassland conditions. Nevertheless, surface water such as streams and water holes were present. It is likely that early hominids were dependent on them, since hominids, like many other mammals, need to drink water at least once a day.

The diet of A. *africanus* has been hypothesized on the basis of dental morphology to be carnivorous or omnivorous. Excavations from Omo indicate that A. *africanus* individuals occur in similar numbers to large mammalian carnivores. If confirmed by further taphonomic studies, this would indicate that these early hominids and carnivores occupied similar positions in the food chain.

A number of sites are known where animal and gracile australopithecine bones are intermingled, but there are no definitely known animal remains from australopithecine meals. Dart at one time considered damaged baboon crania from South Africa to represent australopithecine prey (1957a). The damage to these skulls is now thought to have been caused by rock fragments pushing against the surface of the bone while the skulls were being buried in the consolidating cave breccia.

Gracile australopithecines seemed to have died quite young. It is possible to reconstruct age at death by the wear of the teeth— older individuals show much more wear on their teeth than do younger individuals. A study by Mann showed the mean age at death to be a modern-human equivalent age of 22 years (1973). Ages at death for hominids at Omo show a similar result. Considering the small size of these early hominids and their consequent faster maturation rates, they would have been biologically older at 22 years of age than modern humans. Beynon and Dean (1988) estimate from dental enamel histology that australopithecines would have matured significantly earlier than modern humans, in a pattern of growth reminiscent of modern apes.

Stone tools have never been clearly associated with fossil remains of the gracile australopithecines. Artifacts are lacking at Laetoli, the early portions of the Omo sequence, Makapansgat, Sterkfontein, and Taung. Although stone tools have been reported from Hadar at a possible age of 2.6 million years B.P., they are not associated with the hominid fossils and may be of uncertain stratigraphic provenience. It thus appears unlikely that gracile australopithecines made and used stone tools.

Dart originally suggested that bone fragments found in Makapansgat were in fact tools that australopithecines had used to kill, dismember, and eat animal prey. Dart named the *osteodontokeratic* tool culture on the basis of this evidence, since the supposed tools consisted of bone (*osteo-*), tooth (*-donto-*) and horn (*-keratic*) (1957b). Recent research has shown that much of the damage on fossil bones can be attributed to chewing by hyenas and other carnivores, but some wear and breakage on fossils from the cave of Swartkrans has been ascribed anew to hominid activity by C. K. Brain (Brain et al., 1988). On this basis then it is possible to suggest that early australopithecines were tool makers, although their use of stone for this purpose remains undemonstrated.

An early belief that fire had been utilized at the australopithecine stage prompted Dart to name one specimen *Australopithecus prometheus*, after the figure in Greek mythology who brought fire to earth (1948). K. P. Oakley of the British Museum later demonstrated that what Dart had taken to be carbon, and thus the remains of fire, at Makapansgat was actually a naturally formed manganese oxide deposit (1961). However, recent work by Andrew Sillen and C. K. Brain (1988) suggests that there may have been some fire use at Sterkfontein during the time of *Homo*.

Nonhuman Primate Studies and Early Hominid Behavior

From comparative primate studies it has been shown that important aspects of social structure vary in response to the environmental conditions in which they occur. Group size usually increases and multi-male groups generally occur among primates living in savanna (grassland) or savanna-woodland areas. This suggests relatively large multi-male social groups for the early australopithecines.

Specific analogies have also been drawn between certain primates, particularly the baboon and chimpanzee, and A. africanus. A major contribution in this area has been the demonstration by Jane Goodall that chimpanzees fashion and use tools in obtaining food (1986). (Stones are occasionally used by both chimpanzees and baboons in aggressive throwing or in breaking open hard objects, but they are never purposely chipped.) *A. africanus* can be expected to have shown at least an equivalent degree of tool use.

Chimpanzee groupings are also affected in a predictable way by a change from forest to savanna-woodland, in a manner that mirrors major ecological and behavioral change in hominid evolution. Several primatologists have shown that meat eating and predation, as well as bipedal walking, occur more frequently in chimpanzees that live in more open vegetation habitats, such as forest fringe environments. Early proto-hominids likely possessed such behavioral flexibility which allowed them to adapt from the typical primate habitat, the forest, to a savanna environment in a manner similar to the modern chimpanzee. Among other implications, these findings show that meat eating in hominids has a primate behavioral analogue and did not have to evolve *de novo*.

Ethological studies of large mammalian carnivores, such as lions, spotted hyenas and African hunting dogs, also have provided important clues to australopithecine behavior. A landmark study by George Schaller and G. Lowther showed the relevance of carnivore ecology in the Serengeti Park of Tanzania to early hominid evolution (1969). If australopithecines hunted in social groups, their societies could be expected to resemble unrelated social carnivores more closely than primates, who do not engage regularly in predatory activities. These authors pointed out that early pre-stone-tool-using hominids, such as *A. africanus*, probably made their living by both hunting small animals and scavenging, as do other carnivores. Schaller and Lowther themselves simulated an opportunistic hunting-scavenging australopithecine lifestyle in the Serengeti. They found that, if supplemented by available vegetable food, a day's catch could sustain a small group.

Additional analogies have since been drawn between early hominids and carnivores. For example, early hominids may have had large defended territories, a marked degree of cooperation among adults in food getting, and sharing food with young or infirm members of the group, as do modern social carnivores of Africa. These comparisons underline the fact that while australopithecines arose from a primate base, selection for a social opportunistic and omnivorous way of life significantly altered the trajectory of hominid evolution.

The Robust Australopithecines

The robust australopithecines were a group of hominids which are known to have lived in eastern and southern Africa from 2.5 million years ago to about 1.0 million years ago. Their presence has not yet been confirmed outside Africa. They were specialized para-human creatures, apparently not in modern humans' direct ancestry, yet they coexisted with early members of the genus *Homo*. The causes of both the evolutionary origin and eventual extinction of the robust australopithecines remain obscure (Grine, 1988).

Much of the characteristic robust australopithecine cranial

morphology is related to a specialization for heavy mastication. One student of these hominids has termed them "chewing machines." Characteristic morphology includes a sagittal crest along the midline of the skull and heavy cheek bones (zygomatic arches) to support large muscles of mastication. The face is characteristically dished as in australopithecines generally, but is more heavily constructed than in the gracile species to withstand the forces generated in chewing. To lighten the weight of the head, large areas of the skull have developed internal air cells inside the bone. The East African species, *A. boisei*, and the South African *A. robustus* seem similar in all these respects except that the former appears to be larger and more robust.

From the known cranial endocasts, robust australopithecines apparently possessed brains of an absolute size close to the earlier gracile australopithecines. However, the robust forms probably were of larger body size and therefore would have had a relatively smaller brain.

The teeth of these hominids are specialized for grinding. The chewing surfaces of the molars are expanded, premolars have also become larger and molarlike in form, and the incisors and canines for cutting and tearing food are quite reduced in size. J. T. Robinson put forth the so-called dietary hypothesis to account for this dental pattern (1969). He proposed that the robust australopithecine was a vegetarian who had to eat large quantities of food to sustain its bulk. The dentition of the smaller gracile australopithecine, on the other hand, with its relatively larger incisors and canines and smaller molar surface areas, was adapted to a more varied diet which included meat. Studies on dental wear now in progress— some using the scanning electron microscope—hopefully will help to test this hypothesis.

The post-crania of robust australopithecines is poorly known. On the basis of extremity bones and pelvic fragments the stature of these hominids had been estimated at between 146 and 165 cm with a weight range of 43.2 to 91 kg, somewhat larger than the gracile australopithecine. The arm may also have been relatively longer. Although some foot bones were interpreted as suggesting a divergent big toe in *A. robustus*, indications from the pelvis, putative lower limb bones, as well as the central placement of the foramen magnum strongly suggest well-developed bipedalism.

Robust Australopithecine Paleoecology and Behavior

The robust australopithecines were originally thought to have been the gorillas of the hominid family—large, heavily built creatures living in forested habitats. Initial paleoecological results from South African caves seemed to support this scenario.

Robust australopithecine cave sites had a higher incidence of angular sand grains, presumably washed quickly into the cave and not scattered and rounded over time by wind action. This was interpreted to indicate relatively wet conditions, or greater rainfall than at present, at the time of deposition.

Studies on the fauna, particularly the antelopes, by Elisabeth Vrba have shown that this interpretation should be modified (1982). Dry-adapted bovids predominate in the robust australopithecine sites, indicating drier, not wetter, conditions. Paleoecological results from the several East African sites in which *A. boisei* is known also show a relatively arid climate. The present best estimate of robust australopithecine adaptation is that of a vegetarian savanna dweller.

A number of intriguing paleoanthropological problems still surround the robust australopithecines. According to the dietary hypothesis one might expect to find fossil deposits from these hominids in relatively large numbers because herbivores are generally more abundant in ecological systems than are carnivores. In fact, they occur in virtually the same percentages as the supposedly omnivorous gracile australopithecines.

Another paleoanthropological dilemma associated with the robust australopithecine is ecological niche separation between this hominid and the coexistent *Homo habilis*. Hominids are generally considered culture-bearing primates, but how could two such species live in the same environment without eventually ecologically excluding the other? Perhaps the skill of tool use was exclusive to *Homo*. Robust australopithecines are found in sites in which Oldowan stone tools are also found and, in some cases, where *Homo* is lacking. Despite the absence of *Homo* fossils from these sites, they may have left their stone tools there. General opinion seems to lean toward robust australopithecines lacking the ability to fashion stone tools but the question is still open.

Robust australopithecine species seem to persist in Africa until shortly after the appearance of the relatively advanced *Homo erectus*. Sherwood L. Washburn has suggested that this culturally endowed early human species was able to ecologically outcompete the robust australopithecine and thus drive the latter to extinction (1975). Another possibility is that as *Homo erectus* increased in size through evolutionary time groups, they required greater food resources and territory, and individuals were able to physically outcompete robust australopithecines.

The African Tinder-Box
The Spark That Ignited Our Cultural Heritage
J. Desmond Clark

J. Desmond Clark (University of California, Berkeley), the preeminent African prehistorian, has devoted the last fifty years to teaching and research and continues to be active on three continents. His major work, The Prehistory of Africa *(1970), is currently being revised and updated. For this volume, Professor Clark presents his view of the evidence for early stone tool use and manufacture and the origins of human culture. His latest venture in Ethiopia (with Tim White) has unearthed remains of some of the oldest humans thus far known and some intriguing evidence for the controlled use of fire.*

I suspect that nowadays few, if any, have seen or handled a tinder-box. It is a box for holding tinder and was usually fitted with a flint and steel to make the spark. The tinder is a highly inflammable material (dried moss, fungus, seed down, or dry powdered wood) used for catching the spark when the flint and steel are struck together. This combination was one of the main ways of starting a fire and was used for obtaining heat or light right up to the time when safety matches were invented in 1855. "The African Tinder-Box" seems to be a fitting title since, as is now generally known, it is in Africa that our earliest hominid ancestors evolved some 4.0 to 5.0 million years ago in the highly favorable, ecologically drier tropical savannas of the continent, rich in plant and animal biomass. It was here that the *Homo* lineage sparked the flame that illuminated the path of invention: from the manufacture of the first simple stone tools made in a regular pattern some 2.5 to 2.0 million years ago, through increasingly more complex technological and social behavior, to the first flowering, after forty thousand years ago,

167

of cultural variability that accompanied the spread of anatomically Modern humans throughout the world. Like the Pyrenean shepherd, or the sailor's slow-burning lighter and smoldering wick, the tinder is always there, and all that is needed to ignite it is a new spark.

Many phenomena, the most significant probably being climatic and environmental change, brought about a readjustment and readaptation on the part of ancestral hominids. They developed new and improved ways of using their old resources and introduced new ones. They moved into new ecosystems and, at around 1.0 million years ago, they spilled out of their ancestral continent and began to occupy Eurasia. Our biological evolution took the path of bipedal locomotion which freed the forelimbs for the use, and later, for the manufacture of tools. This, in turn, led to our unique and complex brain. Bear in mind, however, that our increased brain size is the result of a long and complex process of interaction with many other organisms that shared the early hominid habitat. It was a result also of a feedback relationship between the brain, developing intelligence, and communication on the one hand, along with changing technology and improved resource use on the other. Although tools alone did not make us what we are, without galloping technological output, we would not be who we are today. So, what concerns us here is tools, their uses, and the behavior they engendered.

Humankind may not be the only toolmaker and user: birds, otters, and apes, for example, use tools. Rather, it is our unique ability to develop culture and increasingly more complex social organization and behavior, supported by technology, that has made us distinctly human. For example, creatures with small teeth and no claws could not regularly eat meat (other than the smallest animals) without a tool to skin, butcher, and cut it up. This, I believe, is the most significant adaptation that started hominid stone tool use. I will present some of the archaeological methods used in the analysis of stone tools and early hominid lifeways, describe some of the problems faced in reconstructing evolutionary history, and present the current evidence.

Methodologies

Artifacts may be found in two kinds of contexts: primary, meaning minimally disturbed situations, and secondary, meaning situations to which they have been displaced from the original location of deposition, usually through natural causes. Only in the late 1940s or early 1950s was it recognized that, in Africa,

Paleolithic/Earlier Stone Age artifacts were not always located in secondary contexts, such as, in river gravels or glacial outwash deposits, as is the case in much of Europe. Rather, many artifacts are found in primary contexts, such as in association with paleosols, fine-grained fluvial and lake sediments. Prehistorians in Africa were able to identify well-stratified concentrations of artifacts and bone, in fresh condition, lying on old land surfaces (the Rift Valley of Africa, for example). In addition, circumstances are available showing that these artifacts are often only minimally disturbed before being sealed off by overlying sediments. It therefore became possible to excavate part, or the whole, of a Plio-Pleistocene archaeological concentration between 1.0 and 2.0 million years old, and to study the relationships of the different kinds of artifacts to one another, as well as to fractured bone or, more rarely, to plant material, with which some of these are associated. This realization followed from Louis and Mary Leakey's work at Olorgesailie and my studies at Kalambo Falls in the 1940s and 1950s. Together, our work demonstrated that distribution patterns (or examination of the entire context of a site) provide valuable insight into and understanding of the *activities* of the hominids who made the tools, rather than only an understanding of the tools themselves. Thus, research efforts began to be both more scientific and holistic.

Since this realization, all worked stone from an excavation is plotted and retained so that it can be analyzed and quantified, rather than only the retouched tools, as in the past. Broken bones are now examined as possible food waste refuse rather than as only deceased animal species. A more reliable chronology has been established and attention given to activity residue patterns through what is now called horizontal excavation. Through these collective methodologies, we can now better understand the history of a site and speculate on the behaviors the residues reflect. The best kind of site from which to draw information is the short-term, single context occupations, which is usually in the open rather than in caves. For example, while the australopithecine cave sites are excellent for preserving bone, the context leaves little in way of evidence about the behavior of these man-apes. Without an archaeological context, we have only morphology from which to deduce behavior.

In 1959, Mary Leakey found the first early man fossil from East Africa at Olduvai Gorge, *Australopithecus boisei*. She made her discovery by actually crawling, rather than walking, over the ground. This fossil is reliably dated at around 1.75 million years old, and its discovery began the intensified research of the last thirty years in the extremely rich African Rift System. It was a most exciting time to be working. Also, from the early 1960s, interdisciplinary and international teams of scientists—geologists, palynologists, and paleontologists—began working with archaeologists in the field to aid in the interpretation and reconstruction of past events. It is this coordinated teamwork between natural, physical, and social scientists and the pooling of the results (on, for example, micro-stratigraphic sequences and the identification of taphonomic agencies that accumulate and disperse bones and tools) that have made research into Plio-Pleistocene origins so successful over the last twenty years. Through these efforts, prehistorians are now building up an increasingly extensive corpus of information on hominid paleo-environments and behavior. Because of the more detailed knowledge that we now have of the climates and preferred habitats (with their animal and plant life) we can better understand the way in which early hominids evolved.

As yet, however, we know a little about only one very small part of the early hominid tropical world—the African Great Rift Valley. The paleo-climate and biotic data are slowly accumulating, but the archaeologist must still bring all these results together and interpret them in behavioral terms. Some of the approaches and methods used today that are able to develop understanding of the meaning of the archaeological residues in their context are discussed below. These new studies—called "actualistic" from observance of phenomena and processes in action today—are helping toward a more realistic understanding of site residues and so to a more critical reconstruction of early hominid behavior. For example, at one time, a concentration of artifacts and broken bones was regarded invariably as being the work of hominids, but we can now see that there may have been several other agencies at work that were also responsible. We must be able now to identify the characteristic evidences of each of these and show how specifically hominid activities can be identified and isolated.

Streams may concentrate artifacts and bone, or tip them on edge, and a winnowing effect can remove all the small pieces so as to distort what is left. However, micro-stratigraphic study of the sediments and comparison of them with those of present-day stream and flume experiments make it possible to estimate the extent and nature of the disturbance and so to reconstruct the processes that were at work during the history of site formation (Schick, 1986). It can now be seen that many large concentrations have resulted from the accumulating action of streams.

Again, strong winds, especially in arid environments, remove the fine-grained sediments and let down artifacts and bone contained within them onto a lower and resistant surface. This action also causes fortuitous concentrations from successive horizons, as well as wind polishing of artifacts (Clark et al., 1973). Concentrations on old, buried, soil horizons have usually suffered the least rearrangement, and well-preserved skeletal remains of a single large animal in paleosol contexts have, when excavated, been shown to have a number of stone artifacts in association and to be indicative of butchery (Leakey, M.D., 1971:64-66, 85-86). Microstratigraphic studies of fine-grained sediments at early hominid activity sites make it possible to reconstruct the topography and paleo-geography of a site by analogy with those of the present, and a comparison of the evidence from a number of sites can suggest which kinds of habitats were most preferred by the hominids.

Such actualistic studies enable us to identify the sequence of events and the agencies at a site and determine the site's nature and environment. Favored habitats of early hominids were in the forest along streams and lake shores, sometimes on a sand or gravel bar. This is well illustrated in a conjectural reconstruction of a hippo-butchery and food-sharing site dated to around 1.8 million years at Koobi Fora on the northeast of Lake Turkana that the late Glynn Isaac excavated (Isaac et al., 1976:538-41).

Study of the artifacts and other remains of human activity areas are yielding a wealth of information. For example, the experimental replication of artifact sets, such as those of the oldest (Oldowan) hominid toolkits, by Nicholas Toth (1982) show that the resulting forms are the natural product of the flaking process, often stemming from the shape of the original cobble, instead of from a preconceived mental template. The only artifacts that are not a direct product of the reduction process are small, minimally modified or retouched pieces; however, these are rare, except at Olduvai. Many flakes were produced and these are thought to have been the desired end products for use as knives and scrapers to process meat and other materials. (Incidentally, this study by Toth [1985] showed that most Oldowan stoneworkers were right-handed. If the stone to be flaked is struck with a hammerstone held in the right hand, the left hand, which is holding the cobble being struck, is turned toward the body as each flake is removed. When these flakes are refitted to the core, the sequence is from right to left. With a left-handed person the order is from left to right.)

Refitting studies, that is, the replacing of the pieces removed in the process of making of a stone tool, not only show the technique of manufacture, but also several other things. For example, when a plot (graph) is made of the pieces removed (the debitage) as they lie on the horizontally excavated surface, it is possible to identify flaking areas, areas for processing meat and bone, and areas for other special activities. Refitting studies also demonstrate that early hominids carried stones from place to place. For example, at the 1.5-million-year-old site of FxJj50 at Koobi Fora, it was found that several refitting flakes, when compared with an experimental set made by Toth, were missing the original core or chopper. In such instances, we assume that it was carried away by the hominids to be used elsewhere.

Identifying the source of the rocks used for making stone artifacts reveals that the early hominids were very mobile. Often, these sources are several kilometers from the site. Identifying the source also shows that hominids consciously selected what rocks they used. For instance, at Olduvai Gorge, the makers of the Oldowan artifacts found in Bed I and lower Bed II preferred a particular lava for cores and choppers. This stone could have been carried at least 2 km onto the site. At other sites, some of the raw material used implies transportation over distances of 8 to 10 km (Hay, 1976:182-83). At most Acheulian sites where bifaces are found, the stone was obtained elsewhere and carried from the source areas to the activity areas. At Olduvai, in later Bed II up to Bed IV times, the floors were situated by streams draining from the highlands into the playa lake basin. These areas are 18 km distant from the source of the original boulders and outcrops from which the flakes were made (Hay, 1976:183-84). Over time at Olduvai Gorge, stone selection is well shown. Lava and quartzite were used for heavy-duty, larger artifacts, while quartz and chert (when available) were used for light-duty tools.

Use-wear analysis, the study of the microscopic signs of wear and polish on the edges of stone tools, also reveals much information. On stone tools from northern Kenya that are 1.5 million years old (Keeley and Toth, 1981), we can now identify the microscopic signs of use for cutting meat or soft plant material or wood. Also, there is a claim that blood proteins, which are very stable, have been identified on the edges of two tools from Iran that are dated at 100,000 years ago (Loy, 1986).

Researchers have been studying the distribution patterns of bones and the way in which they are broken to extract marrow. The patterns of the fractures and the kinds of modifications seen on the outer, cortical surface of bone fragments make it possible to distinguish the work of the different agencies that have produced these marks. Comparison of what large carnivores do with bones show that lions, leopards, and hyenas, for example, each leave different evidence of their activities. Studies of the ways these large predators in the wild deal with a carcass and bones and studies of hyenas (*Crocuta crocuta*) in captivity have produced clear hallmarks showing which animal is involved (Blumenschine, 1986). Different modification patterns are observable in situations where present-day hunter-gatherers (Hadza or Ova-Tjimba, for example) process a carcass and break up bone for the marrow. Breaking bone with a hammerstone will produce comminuted, depressed fractures, usually to detach the particular ends of long bones and, in breaking the shaft, bone flakes, usually fan shaped, are produced (Potts, 1988). Incidentally, bone marrow is a very important source of fat for foragers, since most wild animals hunted have very little fatty tissue. So ethnographic analogy and controlled experiments in "bone bashing" make it possible to distinguish hominid activity from carnivore activity in archaeological bone and, in a complete prehistoric bone assemblage, to determine what percentage may have been scavenged from carnivore kills as distinguished from what may have been obtained by hunting.

Cut marks on bone that result from the use of a stone knife, usually to remove meat or the periosteum (cartilage-like investment of bone), show butchery practices and are good evidence of hominid activity. Because surface markings made by carnivore teeth are easily distinguishable from cuts made by a stone knife, the sequence of events of a single bone is readily determinable. One such example is a bone fragment from Bed I at Olduvai Gorge, which has hominid cut marks overlain by carnivore tooth marks (Potts and Shipman, 1981). Another example is an archaic *Homo sapiens* cranium, about 500,000 years old, from Bodo in the Afar Rift of Ethiopia, which has cut marks on the face and forehead, showing that the flesh had been intentionally removed with stone knives (White, 1985).

Especially important are studies of present-day carnivore and scavenger behavior: that of lions, leopards, hyenas and others. These studies aid in the reconstruction of the habitats and prey-disposal behavior of extinct forms (Blumenschine, 1986). Through examination of this information, we can determine how much of early hominid diet came from scavenging and how much came from hunting. At present, the consensus of opinion about late Middle to early Upper Pleistocene contexts is that most of the remains of larger game animals came from scavenging and that the remains of smaller- to medium-sized animals came from hunting. It has been suggested that *Homo habilis* got its main source of animal protein from scavenging, although a strong possibility exists that its animal protein could have been obtained from hunted small game, similar to the pattern exhibited by present-day chimpanzees.

The importance of plant foods in the diet of early hunter-gatherers is much more difficult to estimate because only occasionally are plant remains preserved, as are carbonized fruits and nuts at the water-logged Acheulian sites at Kalambo Falls in sub-Saharan Africa. However, micro-stratigraphy and analysis of fossil pollens

from the sediments make it possible to suggest present-day analogies of plant communities growing in comparable environments
and so to conjecture what edible plants might have been available
and thus used in the past. Grindstones are good evidence of plant
foods, and they have now begun to be found in much earlier
contexts than was expected, in the Middle Paleolithic/Middle Stone
Age 100,000 to 50,000 years ago (Clark, 1988:299). From the wear
patterns on the occlusal surfaces of the molar teeth of the hominids,
it would also seem that hard seeds and nuts were an important part
of the diet of the robust australopithecines whereas *Homo habilis*
and *Homo erectus* appear to have been more omnivorous (Walker,
1981; Grine, 1986).

Present-day analogies from primatological field studies and
hunter-gatherer ethnographic works offer another method to obtain
a more complete model for early hominid behavior. Evidence from
great ape behavior provides behavioral models at the early end of
the evolutionary spectrum. For example, models for *Homo habilis*,
the first stone toolmaker, can be constructed based on the
toolmaking behaviors of the chimpanzees (Goodall, 1986). When
we come to early Moderns, we now make use of evidence from the
study of present-day hunter-gatherers. This must be done with
considerable caution, however, because socially and economically,
these present-day peoples have undergone long contact with more
complex societies, pastoralists, and farmers—contact which has
introduced new technologies that often alter entire ways of
existence. Nonetheless, the basic patterns of the hunting and
gathering lifestyle remain and provide valuable insights to the
archaeologist for reconstructing the possible way of life of peoples
in the less-remote past, especially if continuity can be
demonstrated. Where early Modern humans are concerned,
analogies with existing foragers cannot be clearly demonstrated,
and this is even more the case when it comes to archaic humans
for whom no present-day analogies exist. Continued, critical use
of these and other kinds of data, in particular that of the artifacts
and fossils in full context, is nevertheless the basis for the models
that can now be produced to suggest how the unique human
intellectual and behavioral pattern evolved.

These are but some of the ways in which evidence from the
present day is being used to interpret the past. Such actualistic
studies are providing a new and more realistic understanding of the
surviving prehistoric record. We can now begin to identify some
of the things the early hominids were doing, or not doing, thereby
providing the hypotheses on which models can be built and tested.
Present-day analogies, however, can never give any absolute
certainty of interpretation, as there is always more than one possible

explanation for a set of buried data. They do, however, enable us to appreciate better the range of possibilities so that, in testing these models, we can focus our inquiries on eliminating some of the alternatives and so more closely approach the true explanation.

Problems

Actualistic studies are providing a clearer understanding and a greater range of possibilities to some classic inquiries. For example, one question that has long perplexed paleoanthropologists is when hominids first began to use fire. *Homo erectus* and other archaic hominids appear to have been fire users, since some Middle and early Late Pleistocene occupation sites yield evidence of hearths, such as at Zhoukoudian in northern China and the Cave of Hearths and Kalambo Falls in sub-Saharan Africa. Although fire use is associated with *Homo erectus*, *fire making* was probably an invention of anatomically Modern humans sometime after 40,000 years ago. At older sites in East and South Africa and Ethiopia where charcoals are not preserved, studies have been made of reddened earth patches and discolored stones in contexts with evidence of hominid activity, dating back in East Africa and Ethiopia to 1.0 million years or more (Clark and Harris, 1985).

The use of archaeomagnetic analysis of burnt samples can reveal the temperatures at which the earth and/or stones were heated, that of a brushfire being lower than that of a campfire (Barbetti, 1986). SEM (scanning electron microscopy) studies of bone and tooth enamel by Pat Shipman have shown that color and molecular structure of cortical bone change at different temperatures. This makes it possible to show, not only whether bone has been burned, but also, the temperature of the fire. When fire temperatures are around 420 degrees Fahrenheit we know that the hominids were roasting their meat, as has been shown that they did with an eleven-thousand-year-old mammoth from Michigan (Shipman et al., 1984).

Another major problem to be resolved is how long an activity site may have been occupied and/or how many times it was reoccupied before final burial. Attempts have been made to do this by comparing the degrees of weathering of bone on present-day landscapes with those on a prehistoric activity site. Several reoccupations appear to have taken place at one or more of the sites in Bed I at Olduvai, for example (Potts, 1986).

Evidence

The earliest hominid grade associated with stone tools and broken bones is *Homo habilis*. *Homo habilis* is found in both East and

South Africa and is dated to around 2.4 to 2.0 million years. They had a larger brain than the earlier australopithecines, undoubtedly the result of both increased protein in the diet and the means for developing strategies to procure animal meat. Interestingly, the increase in brain size occurs at approximately the same time as the appearance of the first stone tools. Stone flakes (chips struck off from a core) make efficient knives and hammerstones (usually unmodified river cobbles) and are useful for breaking bones in order to extract marrow. These rather crude stone tools associated with *Homo habilis* are referred to as the Oldowan tradition.

Work on Lower and Middle Pleistocene sites on the southeast plateau and the Afar Rift in Ethiopia have given new insight into the earlier Oldowan tool tradition and the later, Acheulian, biface tradition. The Oldowan artifacts and bone lie on a single horizon, and the sites are all quite small in area. Oldowan sites are less extensive than Acheulian sites and the artifacts are all relatively small, while the range of forms is restricted, and one assemblage is very like another. On the other hand, the Acheulian biface tradition (ca. 1.5 million to 200,000 years ago) is innovative and extensive. The area covered by an assemblage is larger, more sites and artifacts exist, and the range of tool forms increases. Several stratified horizons are often present at a site, which indicates reoccupation. The biface component (flaking on both sides of a tool) is new and the technology is varied. For example, at Gadeb on the southeast plateau, large flakes struck from lava boulders were often retouched unifacially (on one side only). At Chorora (Arba) at the southwest end of the Rift, rhyolite was used instead of lava, and the proto-Levallois method (a technique of core preparation for removing a single large flake from the pre-formed upper face) was used to produce the flakes from which the bifaces (handaxes and cleavers) were made. In the Middle Awash region of the Afar, large flakes were removed from very large boulders of lava, a significant proportion of which are called Kombewa (or Janus) flakes. These are flakes with a bulb of percussion and flake surface on both faces. These were then used for biface manufacture. A point of interest at Gadeb is that four handaxes made of obsidian could have come only from the Ethiopian Rift, which is about 100 km distant, since no obsidian sources are known from any closer (Clark, 1987). This indicates that the hominids at this time (1.0 million years ago) ranged over long distances.

There are no artifacts associated with the early australopithecines, *A. afarensis* (ca. 3.75 to 3.3 million years) from Hadar and Laetoli, and *A. africanus* from South Africa. This is somewhat surprising in that there is no doubt about their upright stance and, anatomically, their hands could have manipulated tools.

A robust australopithecine form is present by 2.5 million years in East Africa, and another a little later than 2.0 million years in Southern Africa. These persisted until approximately 1.0 million years ago and overlap with the time that *Homo habilis* existed in Africa until it was replaced by *Homo erectus*, 1.5 million years ago. There has been a suggestion that *A. robustus* made some of the tools found at the cave of Swartkrans in the Transvaal (South Africa). However, since the more evolved hominid *Homo erectus* was also present and the dating very uncertain, this suggestion remains speculative (Brain et al., 1988).

The oldest fossils of the *Homo erectus* grade are the Nariokatome skeleton from West Turkana dated to 1.6 million years, and two crania from Koobi Fora dated only a little younger, which strongly indicates this grade evolved in Africa. The chronology also confirms that the transition from *Homo habilis* to *Homo erectus* was rapid, taking only a few thousand years. The swiftness of the change must have been the result of drastic alterations in nutrition and social organization. *Homo erectus* is the first large hominid with stature like our own and his technology shows similar significant changes with the appearance of the first large, standardized tools: handaxes and cleavers. These tools become a regular addition to the continuing "chopper"/flake-tool tradition from about 1.5 million years ago. While Oldowan artifacts are made from fist-sized or smaller cobbles and stone chunks, the new techno-complex employed the use of larger flakes from which to make the bifaces. These tools are found in sites with much the same kind of paleo-geographical setting as those of the Oldowan tradition, but in a much greater diversity of both sites and forms of tools. They tend to concentrate in places where boulder-sized raw materials occur and exhibit a greater variety and number of retouched tool forms. Evidence from Kalambo Falls in Zambia and Olorgesailie in Kenya, where several stratified activity horizons exist, suggest that these sites were reoccupied over long periods of time.

By about 1.0 million years ago, hominids began to move out of Africa and into West and Southeast Asia, the Far East, and Europe. The standardized biface tradition is found in some of these regions but not in others (Klein, 1989:212-16). Many of the assemblages give the impression of "sameness," as if the many different kinds of resources were all being exploited in much the same way and at the same low level of intensity in spite of the diversity of the biota. In the Far East and the tropical forests of Southeast Asia, the Acheulian biface is not found and the toolkit still resembles the Oldowan "chopper" and flake-tool tradition. This could be, in part, because perishable materials, such as bamboo, wood, bark and fibers, were being used for tools and equipment, in which case, the stone artifacts represent only the surviving elements in the material culture.

Recently, a claim has been made that certain stone artifacts from Siwalik sediments in northern Pakistan are 2.0 million years old (Dennell et al., 1988) although no evidence of australopithecines or hominids earlier than 1.0 million years have been found there; clearly further evidence is needed. *Homo erectus* in Africa differs morphologically from the Asian forms, which may have acquired their specialized characteristics from long isolation in eastern Asia, although some researchers suggest that *Homo erectus* might have evolved independently in Asia.

In Africa, from approximately 0.5 million years onward, the fossils show an early development from *Homo erectus* into what are called early archaic *Homo sapiens*. The crania of these fossils combine an expanded brain case and other *sapiens* features with a still-massive face and supraorbital bossing that suggests strong masticatory muscles remained necessary. Samples of these crania come from Broken Hill (Kabwe), Bodo, Saldanha, Hathnora, Ndutu and Sale (Brauer, 1989; Stringer and Andrews, 1988). At sites where artifacts are found in association with the fossils, they belong to the Middle Pleistocene biface tradition, whether from Africa or India (Hathnora). A range of variability exists in the shape of the handaxes and cleavers, spatially and temporarily; however, other artifacts, such as heavy- and light-duty scrapers, other light-duty tools, spheroids, and so on, are all very much the same, regardless of the raw material. This narrow range of technological and typological forms seen in the biface tradition implies that Middle Pleistocene hominids were practicing similar adaptive strategies of food procurement and use of resources. Given the wide range of ecosystems occupied, and plant and animal resources exploited, the level of extraction efficiency must have been both low and unspecialized. No doubt, if communication had been more efficient, it would have been reflected (as in the Upper Paleolithic) in greater complexity, diversity, and regional specialization of material culture.

As with the fossils, the archaeology in Africa also shows a smooth transition from the Lower to the Middle Paleolithic (ca. 200,000 to 100,000 years ago). The only significant changes in the stone toolkit are the disappearance of the large bifaces (handaxes and cleavers) and the refinement of the prepared core technique for obtaining thin flakes and blades. A major technological advance was the intentional hafting of stone parts of tools to form simple composite tools and weapons. The "tang," an innovation from northern Africa, provides incontrovertible proof of this technological advance. The tang is a characteristic tool of the Aterian industries of northwest

Africa and the Sahara, made to fit into a hollow sleeve and then secured with mastic. Elsewhere, a sign of hafting is seen in the thinning of the butt of a tool. The practice of hafting made for lighter equipment and also made possible a greater variety of tools, according to how the working parts were mounted. Various kinds of "points" exist from this time and may have served as knives or spear tips made with both a cutting and/or piercing head. The Middle Paleolithic/Middle Stone Age provides, then, the earliest evidence for regional diversity and specialization.

The first anatomically Modern human fossils discovered outside of Africa are from the Levant (the geographic area bordering the eastern Mediterranean) and are dated from 90,000 to 100,000 years ago. The fossils come from the caves of Qafzeh and Skhūl in Israel and are burials. These early Moderns are thought to have entered Western Asia from North Africa (Vandermeersch, 1989). Neanderthals, descendants of the archaic *Homo sapiens* population that first entered Europe, seem to have moved into the Levant from Europe about 60,000 to 70,000 years ago. It has been suggested that this migration occurred because of the southward advance of the glaciers (Valladas et al., 1987). By 40,000 B.P., however, the Neanderthals disappeared, it seems, without hybridizing with early Moderns, who may, in fact, have returned to Africa during the thirty-thousand-year interval when the Neanderthals resided in the Near East. Although there is some vehement denial, it seems more likely that none of the non-African paleoanthropic populations of the Old World (archaic *Homo sapiens* and Neanderthals) made any significant contribution to the Modern (*Homo sapiens sapiens*) gene pool.

Four years ago, human paleontologists and prehistorians were galvanized into action by an announcement made by molecular biologists (Cann et al., 1987). They stated that the mitochondrial DNA of people alive today, assuming it changed at a constant rate, could be used, by counting the number of genetic mutations, to show when our own species began. Also, by comparing the number of mutations in people of today, the closeness to or distance from the first Moderns could be determined. These results showed that the greatest amount of variation and so of antiquity was to be seen in Africa south of the Sahara and that, therefore, anatomically Modern humans (*Homo sapiens sapiens*) originated in Africa sometime, it is estimated, between 200,000 to 100,000 years ago.

New, critical reevaluation of the fossil record and the availability of various new dating techniques appear also to confirm that the gradual expansion of Modern characteristics occurs through archaic *Homo sapiens* into fully Modern fossils from sealed sites in Africa that are associated with a Middle-Paleolithic/Middle-Stone-Age technology. Crania from north Africa (e.g., Dar-es-Soltan, Tamara); from East Africa (Omo I, Ngaloba); and South Africa (e.g., Border Cave, Klassies River Mouth) in contexts that are both in the open and in caves, are dated to between 150,000 to 50,000 years ago and are all characteristically early Moderns.

More recently, the statistical counting procedures used for calculating mitochondrial DNA variability and chronology have been challenged. Clearly, more research is essential to confirm or refute the claims for an African origin. However, the work on nuclear DNA is not in question and similarly shows the greatest variability and antiquity to be in Africa, although it can contribute nothing to dating the first appearance of the Modern genome.

From 40,000 B.P. onward, Modern populations exploded into Western Asia, Europe, Southeast Asia, and the Far East. Anatomically Modern humans were in Australia by 40,000 years ago, or perhaps even earlier, having crossed 100 km of open water to arrive. In Melanesia, stone tool makers are present by 20,000 to 30,000 B.P. and, around 12,000 B.P. or earlier, they crossed the Bering land bridge and spread out throughout the New World. New technological advances characterize these newly exploited regions. It is clear that the technology of Modern humans, particularly during the Upper Paleolithic/Later Stone Age, takes a quantum leap forward. In the northern latitudes of Eurasia, tools of bone, ivory, and antler are added to the stone toolkit. In addition to technological advance, there is a whole range of evidence that indicates abstract thought: burials with grave goods, ornamentation, and artistic expression all characterize the Upper Paleolithic. In short, the ancestors of our own kind had "arrived" and were individuals with thought processes similar to our own.

What was the catalyst that made Modern humans so much in advance of Neanderthals or other archaic humans? Personally, as I said as long ago as 1970, "The achievement of awareness and integration, as with the transmission of knowledge, is in the main through speech" (1970:146), and I believe that a fully developed language system, such as our own, is the secret of their success. Recently, geneticists (Cavalli-Sforza et al., 1988) have shown that genetic grouping and development in language evolution could have been the determining factor in the rapid expansion of Modern humans throughout the world. The archaeological evidence, although circumstantial, all points to its having been spoken syntactic language that struck the spark.

The ramifications of our family tree are, however, still very far from being clearly understood, as are the behavioral patterns that are the outcome of our biological and cultural evolution. The discovery of a new fossil or some new archaeological data can significantly alter the way the evidence is interpreted. The 2.5-million-year-old "Black Skull" (robust australopithecine) in northern Kenya, for example, added a new lineage branch (Walker et al., 1986). Again, identification of criteria for distinguishing scavenging from hunting and for determining meat and other resource processing from use-polishes on stone tool edges have all contributed to reinterpretations of the data. Clearly, the vital need is for more of the hard evidence from stones and bones in their paleogeographic contexts. Archaeologists are now also beginning to look at their material in terms of the way a prehistoric group used its space and the resources this contained, so that activity places are no longer regarded simply as a number of self-contained entities, but as parts of a dynamic system in which even isolated artifact scatters assume new significance. A population of early hunter-gatherers can be expected to have been at least as nomadic as their present-day counterparts; consequently, the evidence extracted from each site can only show what was going on at that particular site. We are now, therefore, looking to identify the different ways in which a prehistoric group used its resources and the whole terrain it exploited, the better to try to understand the yearly, seasonal, and diurnal patterns of the individuals making up the group. Essential for this is study of the present-day ecology and also a knowledge of the paleogeography of a particular area and its plant and animal communities, in other words, the resources that form the subsistence base.

Fresh possibilities are resulting from many new lines of inquiry: remote sensing and resolution of satellite imagery show subsurface topography; deep coring in lakes provides a record of vegetational history and deep-sea cores a record of world temperature changes; stable isotopes of carbon and nitrogen, present in fresh bone, will, if they can be isolated in fossil bone, give exact information on hominid diet. Many more actualistic studies are needed as we replace speculation with proof, and our dating techniques need to be more precise. A reliable time scale will document population expansion; migration will be better understood; and why Africa was where humanity originated, and where, it seems, Modern humans also first appeared in the world may also become clear.

Thus, we need to broaden the search by refining our methodology. But we also need to intensify the search in tropical Asia, where, up to now, much less systematic work has been carried out. If there are, indeed, 2.0-million-year-old tools in south Asia, then there would also have to have been the hominids there to make them. The proof will result only from the thoroughness of future investigations.

We can be confident that these new approaches will provide new information about early hominid behavior and about the sites and the activities that were carried out at them. But the evidence that the occupation sites themselves can provide about expanding intellectual horizons or aggressiveness or communication systems, for example, is much more circumstantial; and if, as seems probable, increase in mental ability and technological complexity is a consequence of a feedback relationship, then the best hope of documenting the evolutionary stages that led up to Modern humans lies, I believe, in closer collaboration and teamwork than has existed up to now between physical anthropologists and archaeologists as well as many other scientists.

It took forty thousand years for humans like ourselves to develop the richness of ethnic and linguistic diversity that we see in the world today. Having now the ability and technology to control the destiny of all creatures on this planet, we must realize that the pace of technical evolution is such that we have only a fleeting moment left in which to work together to establish controls for saving from extinction the plants and animals that have been our companions in our evolutionary past, and, furthermore, by mutual under-standing and appreciation of the lifeways and traditions of our fellow humans, to ensure that they and we may survive to share a rich future together. Understanding the process of our behavioral evolution can help us to develop the vision and means to this end.

IV

FUTURE

There is nothing we do today which will not be done better tomorrow.
—S. L. Washburn (1951).

In 1951, S. L. Washburn presented a paper for the New York Academy of Sciences entitled "The New Physical Anthropology." At the time, physical anthropology was "considered primarily as a technique"; a means for measuring, classifying, and comparing living and fossil specimens. Washburn argued for change: a change of view which would concern itself ". . . with process and with the mechanism of evolutionary change." To facilitate a change of view, he felt it was imperative that physical anthropology begin a collaboration with many disciplines. Of primacy at the time was the "infusion of genetics into paleontology" (1951:298). This would provide evolutionary studies with "a clearly formulated... conceptual scheme" and revolutionize the discipline. However, even before this could be accomplished, a "clearing away [of] a mass of antiquated theories and attitudes" would first be necessary (1951:299). Thus, in the 1950s as well as today, the growth and future of physical anthropology is contingent upon input from various disciplines, new conceptual models, and new attitudes.

The prediction that genetics would ultimately be incorporated into physical anthropology has certainly been born out. Our understanding of evolutionary change through genetics and molecular studies has become integral to physical anthropology. Nuclear and mitochondrial DNA studies have decidedly altered our understanding of our relationship and divergence with the great apes.

In addition to the inclusion of genetics, physical anthropology today regularly includes the input from many other disciplines. Studies from diverse fields such as demography, primatology, biomechanics, geochronology, taphonomy, archaeology, faunal analysis, and paleoecology, to name a few, all keep physical anthropology alive and vigorous. Far from the stagnant quantifying studies of the past, physical anthropology is now dynamic—a process in and of itself.

Perhaps the most significant area that has lead to the rapid expansion of the discipline is the development of new technology. In the 1950s, the primary tool of the physical anthropologist was the caliper! Needless to say, no matter how carefully the measurements were taken, the limitations of such restricted technology is obvious. In direct contrast, the technology of the 1990s is expansive and transformative. The scanning electron microscope allows minute patterns of tooth wear, bone fragmentation, and tool manufacture to be magnified, making analyses unknown before possible. Advanced technology for accurate dating of fossils, artifacts, and sites has literally forced a reevaluation of previous beliefs about our ancestry. We can be assured that reevaluations will continue as new technological innovations are applied to the discipline.

In the final two chapters, the future of physical anthropology is considered from two perspectives: the technological and the intellectual. In "Visualization and Physical Anthropology," Carl Hartwig and Lewis Sadler discuss the potential of computer visual graphics in relation to physical anthropology. In the final chapter, "Education and Evolution, Revisited," Washburn discusses the importance of evolutionary studies of the educational process.

Although the discipline of physical anthropology is entirely different today than in the 1950s, some aspects remain the same. Change within a discipline can only occur when new attitudes, methods, and technologies are examined and embraced. Change and growth is not possible unless the mind remains open. Looking back to Washburn's 1951 paper, the conclusion was as predictive as it remains today: "There is nothing we do today which will not be done better tomorrow."

Visualization and Physical Anthropology

Walter Carl Hartwig and Lewis L. Sadler

In this chapter, Walter Hartwig (University of California, Berkeley) and Lewis Sadler (University of Illinois, Chicago) comment interestingly on new computer techniques that have the potential to generate a new and unique storage facility for anthropological data retrieval and information sharing.

"What we have here is an unsurmountable opportunity."
  Yogi Berra

In any given scientific discipline, priority for research and development funding is given to programs which deliver the optimal combination of cost-effectiveness and public benefit. National defense and medical research logically receive high status both in terms of actual dollars spent and facilities allocated. Near the opposite end of this continuum are fields such as physical anthropology, which offer ostensibly lower-impact "results" or "discoveries." One result of this industrial scale is that as a research group, physical anthropologists usually operate with perpetually outdated technology and without the support of major corporate or public sponsors. An initiative toward the building of collaboratories,[1] by merging the collective resources of the academic field with developments in automated information systems, should bring physical anthropology into step with the interests of more diverse funding sources than it currently accesses.

One of the more active areas of research and development across fields with comparatively open-ended budgets has been visualization technology (McCormick et al., 1987), usually couched in the lexicon of biomedical visualization or computer-aided design and manufacturing (CAD/CAM). Visualization is widely regarded as one of the most important advancements in scientific computing in the last decade (Gore, 1989). Considered here, visualization will refer to the entire process of image acquisition, processing and analysis (see table 1), and it represents a progressive direction for morpological interpretation to take. This review traces the historical applications of visualization in physical anthropology and outlines an agenda for the future that capitalizes on the opportunity for state-of-the-art research as well as the imperative for a cooperative, collaborative, long-term data networking protocol.

Visualization recognizes that far more than half of the human brain is devoted to the processing of images and that visual images are the most efficient way to communicate and understand information. At the same time, the measurement and quantification of the properties of objects are the working bases of science. When information is created we objectify it and embed it in artifacts, books, journals, film, slides, and photographs. However, once converted to a symbol inscribed on a page or film it remains passive and static. The computer, because it can process data interactively, allows the data to be recombined and used in new and historically unprecedented ways. Expanding on this, Paul Gray, president of the Massachusetts Institute of Technology, recently testified before Congress:

> . . . this is a very fundamental change in how science is pursued.
> In fact, nothing like this has happened in three hundred years.
> The scientific method as we know it today was born in the 1600s
> with the mathematics of Descartes and the empirical
> observations of Newton, of course, and the proverbial apple that
> landed on his head. Those are the two basic paths to scientific
> discovery that have been used since that time. Numerical
> simulations represent a . . . new, third path toward scientific
> discovery (Gore, 1989:28).

John Rollwagen, chairman and CEO of Cray Research, Inc, added:

> . . . this is very much the beginning of a cultural change — a
> paradigm shift — to computational science . . . the computational
> approach will move quickly to take its place along side of
> experimentation and theory in the pursuit of new heights of
> science and technology (Gore, 1989:31).

Biomedical image science is destined to have a major impact on research and education in physical anthropology.

Table 1

Biomedical Visualization Terminology

biomedical visualization: The conversion of data to images to take advantage of the human visual system for interpretation and analysis.

CAD/CAM: Computer Aided Design and Computer Aided Manufacturing. This acronym refers to software most commonly used in industrial design product manufacture, principally for part specifications and exact instrumentation.

collaboratory: literally, a cooperative network of laboratories. The ideal collaboratory setting includes and builds upon information exchange possible over high speed data networks and advanced computing.

computed tomography (CT) or computer-aided tomography: A CT scanner uses x-rays to image material densities in volumetric space, thus providing a digital record of an object that can then be viewed independently of the object itself.

data networks: communications channels for digital data transmission. Data networks are often characterized as local area networks (LAN), metropolitan area networks (MAN), and wide area networks (WAN).

digital image archive: a computer database of image data (such as CT data) that is curated or archived at a central location and transmitted over data networks to users for local analysis and study.

digital media: magnetic tape, optical disk, "floppy" disks, compact disk, hard disk, etc.

laser digitizers: ranging devices that triangulate 3D surface data from the coherent light generated by a low power laser.

magnetic digitizers: systems that measure object shape by disturbances in a magnetic field surrounding the object.

pixel: the quantum unit of a computer image; in two dimensions, the unit of resolution is a pixel. In three dimensions, or volumetric space, the unit of resolution is a voxel.

sonic digitizers: systems that triangulate 3D coordinate data from sound waves.

stereophotogrammetry: the process of calculating surface information from two offset (stereo) image views of the same object.

video data acquisition systems (VDAS): These data gathering systems are useful in two-dimensional data collection, where the data are gathered, manipulated and analyzed numerically from pixel locations. The accuracy of VDAS depends on the optical and shape characteristics of the specimens as well as the pixel resolution of the display, or output, devices.

Information Science

Many research problems in physical anthropology and bio-medicine require the same analytical approach despite sometimes radically different databases. The emerging fields of biomedical visualization and health informatics are dedicated to the infrastructure of clinical diagnosis and medical education, and as such have many tools to offer the physical anthropologist. For example, ontogenetic changes in the cranial base and face are as meaningful to the reconstructive surgeon as they are to the comparative primate morphologist, but the data collection systems developed for the former (Marsh and Vannier, 1983; Cutting et al., 1986; Coombes et al., 1990; Yasuda et al., 1990) are rarely used by the latter. Likewise, the entire subdiscipline of materials research in dentistry, which has developed sophisticated methods for imaging and reproducing tooth morphology (Young and Altschuler, 1981; Rekow and Erdman, 1985; Bayne, 1989; Leinfelder et al., 1989), is rarely consulted by physical anthropologists approaching the same methodological problems on modern or fossil primate specimens. For many, if not most, programs in physical anthropology, the financial cost of acquiring this technology is prohibitive. Unfortunately, the intellectual cost of operating in parallel but without it is even greater. Laboratory integration, rather than independence, will minimize both. Collaboratories that share information technology across computer networks will benefit from the exchange of ideas and data without necessarily compromising the research or production of each as an independent entity. Collaboratories are the logical extensions of visualization technology, and the necessary products of information science research and development. Physical anthropology as a field will participate in collaboratory research only if it gets "on-line" with existing engineering and biomedical technology.

Many levels of data integrity exist within any scientific domain. In paleoanthropology, for example, primary (first-order) data are the physical objects collected during fieldwork. Secondary data derive from the initial manipulation of primary data, and may assume a variety of types that are almost always quantified. Further statistical analysis, such as an analysis of variance in some aspect of the secondary data, constitutes tertiary data, and so on, until the information resulting from analysis is completely removed from the empirical data (see figure 1). Several important procedural qualities, such as consensus, replicability and objectivity, are difficult to sustain at the high end of this integrity pyramid due to the concomitant increase in subjective decision making. According to this framework, any expansion of the primary database would

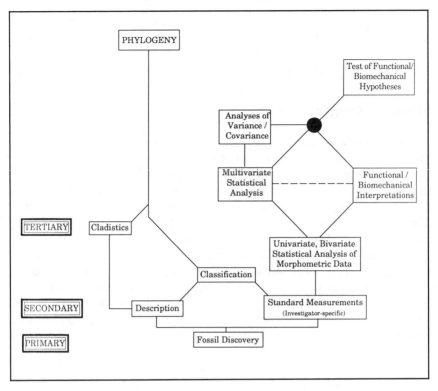

Figure 1. *Flowchart depicting a scheme of data analyses in evolutionary anthropology. A fossil discovery is used as an example of primary data, which actually include a wide variety of physical specimens.*

strengthen the foundation of subsequent analysis and interpretation. Visualization greatly expands the primary database of the morphological sciences by capturing a dimension (depth of field) previously out of reach.

No computer simulation can match the human brain for qualitative assessments, but the latter are greatly supplemented by the quantitative capacity of the former. In addition to bringing a finer degree of resolution to primary data, visualization opens new directions for secondary analysis without in any way compromising the existing infrastructure of conventional morphometrics. The technological and structural specifications of the visualization archive discussed below are designed specifically to mesh with the primary database of physical anthropology.

Analyzing shape within an objective framework continues to be a goal in all morphological sciences. To detail the historical background of morphometrics in physical anthropology would be to detail the fiber of the field itself, so we instead focus on a few historically sequential applications of visualization technology to particular research problems. This is a selective rather than comprehensive review, intended to highlight the progress made in visualization technology and the direction of future research.

Stereophotogrammetry

In its broadest definition, stereophotogrammetry merges two two-dimensional images taken of the same object from different perspectives and creates a field of depth by aligning them according to arbitrary points of registration. The basic technique is widely used in the qualitative assessment of aerial photographs and has been customized to various other fields of study, primarily cartography, radiology, orthopedics, kinematics, and growth and development. In the initial rendering, the data are images only and lack a numerical representation. The images are then digitized to yield sets of coordinates that are merged by one of several available computer algorithms. One of the most popular applications of stereophotogrammetry has been in radiology and orthopedics, where radiographs are substituted for the photographs and three-dimensional data may be obtained on landmarked points through a kinematic sequence. Roentgen stereophotogrammetry of metallic markers implanted in the skeleton has been used to study growth (Alberius et al., 1990), joint kinematics and surgical preplanning (Selvik, 1990). Measurement accuracy is considered greater than that of previous techniques used in longitudinal studies (Alberius et al., 1990) but clearly depends upon the initial accuracy of the radiographs.

Although the technique of taking measurements from a three-dimensional photographic acquisition system has obvious applications to physical anthropology (Herron, 1972; Creel, 1978), it can be considerably more labor intensive and time consuming than most conventional measurement techniques and thus has not proven to be reason alone for moving from two- to three-dimensional approaches. Automating the procedure will help considerably. Recent applications in physical anthropology have taken advantage of the capacity of stereophotogrammetry to visualize small objects, such as the occlusal surface of a tooth (Rekow, 1987; Hartman, 1989), or to model growth (Burke and Hughes-Lawson, 1989). Unlike traditional photogrammetry, roentgen stereophotogram-metry is almost completely automated (though still time consuming) and offers the potential for volumetric data analysis. The prospect of measuring growth longitudinally by tracking a series of custom-implanted tantalum markers seems limited only by the choice of marker location and thoroughness of measurement

registration. As a visualization system, however, stereophotogrammetry is still a relatively cumbersome means of acquiring a digital,
topological database.

Scanning Electron Microscopy

The most fertile area of scanning electron microscopy (SEM)
research in physical anthropology has been the analysis of dental
microwear (see figure 2). The study of diet and the mechanics of
mastication pervades primatology, and microwear analysis has
begun to explore the suggestions of Butler (1952), Mills (1955) and
Dahlberg and Kinzey (1962) that dietary diagnoses could be made
from microscopic wear patterns. No progress was made in this
visualization medium, however, until Walker's (1976) application
of light microscopy on teeth with observational studies of
cercopithecoid feeding behavior.

In the meantime, Boyde (1967, 1971) had demonstrated the vast
potential of SEM for visualizing dental structure, and improved
casting techniques enabled researchers to conduct longitudinal and
museum-based studies that had previously been restricted. In the
last fifteen years, and particularly the last five, SEM analyses have
widely canvassed the field of applied dental research (Shkurkin et
al., 1975; Walker et al., 1978; Puech, 1979, 1981; Ryan, 1979;
Walker, 1981; Gordon, 1982; Peters, 1982; Puech et al., 1983;
Teaford and Walker, 1983a, b; Teaford, 1985, 1988; Biknevicius,
1986; Hildebolt et al., 1986; Teaford and Byrd, 1989; Teaford and
Oyen, 1989a, b, c; Walker and Teaford, 1989). The accumulated
research efforts (reviewed by Teaford [1988]) have raised as many
new questions as answered old ones (Teaford and Oyen 1989c), and
have brought the field of dental micro-morphology to the verge of
a visualization breakthrough.

We agree with Walker and Teaford (1989) that SEM, as a
visualization medium, now requires robust automated analytical
procedures. An initial step in this regard (Grine, 1986) has been
followed by direct application of optical diffraction methods (fourier
transformations) to micrographs (Kay, 1987; Kay and Grine, 1989).
This follows earlier applications of these techniques in other realms
of primate biology (Oxnard, 1973) and is a progressive attempt to
systematize some of the subjectivity that is inevitable in microwear
image analysis. The innovative marriage of optical diffraction
methods and scanning electron microscopy is not without problems
inherent to the nonphotographic nature of the latter, but we regard
the application as a successful harnessing of SEM's visualizing
capacity. As noted by Walker and Teaford (1989), a major advance
in analysis is at hand.

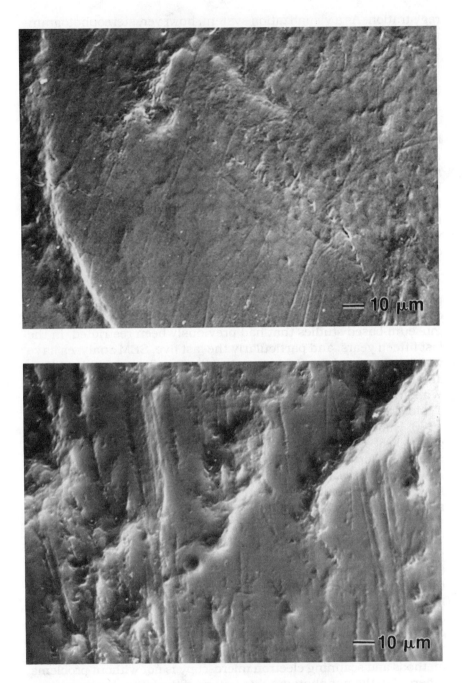

Figure 2. *An SEM photograph of dental microwear in two New World monkeys—*(top) Alouatta; (bottom) Chiropotes—*illustrating the high resolution obtained by SEM imaging.* (Photos courtesy of Mark Teaford.)

Computed Tomography

Computed tomography (CT) scanning has revolutionized clinical diagnosis by providing a noninvasive window into volumetric structure. Although the technology has been operative for almost twenty years, only recently has it been applied to skeletal structure and paleontology research (Tate and Cann, 1982; Conroy and Vannier, 1984, 1987; Borkan et al., 1985; Vannier et al., 1985; Falk et al., 1986; Haubitz et al., 1988; Daegling, 1989; Grine et al., 1989; Le Floch-Prigent, 1989a, b; McGowan, 1989; Conroy et al., 1990; Thomason, 1990; Luo and Ketten, 1991; see Ruff and Leo, 1986, for a useful review of technological specifications and operative procedures). To date, most studies have incorporated computed tomography purely as a means of "seeing" internal structures, rather than as a means of obtaining quantitative data, though some have speculated on this possibility (Conroy and Vannier, 1984, 1987). Early work on the potential for three-dimensional reconstruction from CT data was encouraging and suggested instantaneous, accurate general measurements (Vannier et al., 1985). A direct study of the capacity of CT images to yield accurate measurements brought to light the most critical issue for future applications of this medium:

> It is clear that many people consider 3D CT imaging to be a quantum advancement, but is this the case? The ultimate goal in producing 3D images is to convey information about objects to observers (Hanson, 1985; McCormick et al., 1987). The value of this information depends on its usefulness . . . So far, most of the 3D CT effort has been directed toward making the "pretty pictures" prettier. Terms such as "fidelity," "precision," "quality," and "artifact" have been used as artistic qualifiers rather than as scientific descriptors. Because the terms have not been defined with regard to 3D CT and because statements involving these terms have not been supported with adequate quantitative data, evaluations of image-processing parameters and judgments of image usefulness are of limited value. Moreover, without quantitation, one cannot completely characterize normal and abnormal skull morphology or measure the effects of growth and treatment (Hildebolt et al., 1990:284).

The authors of the study noted that a great deal of image enhancement is required for high-precision results. While image enhancement is necessary for control of bony landmarks such as sutures and foramina, it is becoming less expensive in terms of computer time and actual cost. It should also be noted that the expense of imaging and enhancement is singular and front-ended, not iterative as is the case for conventional data collection techniques, and is ultimately compensated for by the value of the accessible data archive it creates.

The primary issue for anthropologists and paleontologists here is ease and accuracy of measurement. Hildeboldt et al. (1990) performed a landmark study on five adult human skulls comparing three data acquisition techniques: manual caliper, automated digitizer, and CT data collection media, following a study of the former two methods that confirmed their concordance (Hildeboldt and Vannier 1988). They found that several measurements differed by between 1-10 mm across media and concluded that while no statistically significant differences existed, substantive differences precluded uncritical acceptance of CT measurements as substitutes for traditional, caliper-derived data. Room for improvement included landmark location strategies and the length of time required for image enhancement prior to 3D reconstruction.

We suggest that some technical adjustments and a broader protocol attitude will solve most of these shortcomings. In order to improve CT image quality without impractically increasing operative time, we suggest scanning with overlapping rather than contiguous table increments. In the Hildeboldt et al. (1990) study, 2.0 mm wide CT scans were taken at 2.0 mm intervals. A scanning protocol of 1.5 mm slices taken every 1.0 mm yields greater data control but is clearly more expensive computationally (see figure 3).

Also of special note are the nonmedical industrial CT scanners (SMS Computerized Industrial Tomographic Analyzer, for example) that are used for reverse engineering and product verification. These systems utilize a radiation source such as an x-ray tube, an x-ray linear accelerator, or a gamma ray-emitting radioisotope. The

Figure 3. (page 197, top) *This unenhanced image was produced on standard x-ray film by a General Electric 8800 computed tomography scanner by integrating 1.5 mm scan slices taken in overlapping 1.0 mm increments. The specimen is an infant skull (approximately 6 months old) recovered from an archaeological site in California. Even this initial image rendering depicts suture lines, fontanelles, and the interior cranial surface.* (Courtesy of The University of California, San Francisco.)

(page 197, bottom) *The same data are now rotated and hemisected to reveal internal cranial anatomy inaccessible to manual inspection. The extremely rare and fragile nature of this specimen call for visualization techniques that eliminate unnecessary handling and potential damage.* (Courtesy of University of California, San Francisco.)

(page 198) *Posterior view of the upper half of the specimen, imaged from CT data and enhanced on an IBM workstation; the individual bones were aligned on the CT table as a composite that best approximates anatomical position.* (Courtesy of IBM Corporation Palo Alto Scientific Center.)

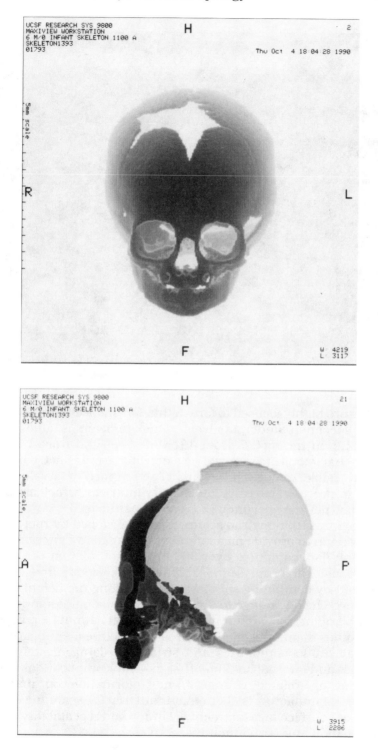

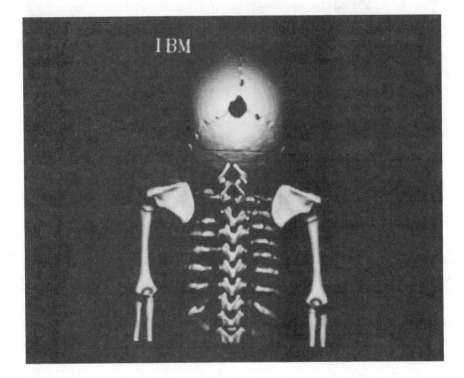

photons are highly focused to form a thin fan-shaped beam directed at the specimen under analysis. The fan-beam is adjustable, depending on the model, from 18-30 degrees and from 1-5 mm thick. The photons are received by a detector array which converts them to visible analog light events. The scanned data values are computer processed to calculate density matrices, which are used to reconstruct the specimen cross-sectional image.

Tomograms in industrial scanners are developed by rotation of the specimen to provide measurements along many interior axes. Projection data computed over 360 degrees of rotation produce a two-dimensional cross section. Three-dimensional images are generated by combining successive scans along the height of the specimen. The advantage to any CT scanner is that it allows data collection of internal anatomical features that cannot be recorded in any other manner. Added features include the identification of potential subtle pathologies and structural damage, bony wall thickness and internal anatomy that may have implications about diet- and age-related changes, and information on material densities. Manufacturers claim that industrial CT scanning equipment has a surface measurement accuracy and repeatability range of 0.001 inch and 0.005 inch respectively.

Technical refinements in image processing are inevitable (Ney et al., 1990; Yasuda et al., 1990), and a large number of software packages for three-dimensional image reconstruction have been developed (Huijsmans et al., 1986; Vannier and Conroy, 1989b). Of greater importance to the realistic manifestation of these developments within the means of basic science research is a broad-scale cooperative protocol for data collection and archive networking. We believe strongly that a program of systematic, image-based information exchange, generated through CT and other visualization media and sponsored by a consortium of corporate and public interests, is the quantum advancement so opportunely available to the morphological sciences and encouraged by the fluorescence of biomedical visualization.

For many physical anthropologists, primary data are less meaningful than the information obtained from them through measurement and quantitative analysis. Measurement techniques and statistical analyses have developed in parallel with improvements in visualization data control. Fundamental questions of shape change require three-dimensional databases for analysis, but the technology for acquiring such data with high accuracy has only recently been developed (Richtsmeier, 1990). Modelling approaches such as finite element scaling analysis and generalized resistant-fit theta-rho analysis (GRFTRA) are already replacing conventional, two-dimensional morphometric approaches in physical anthropology research (Reilly, 1990). The most logical advantage of a three-dimensional visualization system is that it makes available the type of data necessary for sophisticated mathematical analyses without compromising the integrity or accuracy of data used in traditional morphometrics.

Visualization Systems

In essence, the 3D coordinate values of individual surface points only provide information about the geometry of a surface; a complete surface representation also requires topological information (i.e., how each data point is connected to its neighbors). The selection of one scanning procedure over any other should depend not only on how the XYZ coordinates are collected, how accurate the collection may be, and the speed at which they are recorded, but also on the completeness of the topological information. Some data collection procedures can provide complete topology, while others provide partial, or no topology. It is the lack of topology in historical collection methodologies that has severely compromised the usefulness of their data. Any data lacking topological information must be subjected to some mathematical topology generator in order to reestablish the topology from the available geometry. Therefore, the selection of one data collection method over another also involves a conscious decision to use one data processor over another. The following section surveys available visualization systems and their specifications.

Manual Data Acquisition

Measurements made using standard morphometric instruments of physical anthropology include straight-line, angular, and contour (circumferential) data (e.g., sliding and spreading calipers, spanners, bone boards, goniometers, metric tape, and so on). These instruments have been improved somewhat in recent years through the addition of LCD (liquid crystal display) readouts on calipers and direct input to computers or printer devices to speed data acquisition and improve accuracy of recording. No topological information is collected.

3D Data Digitizers

A variety of manual, stylus-type 3D digitizers are commercially available. Some systems utilize translation and rotation-sensing transducers to determine the mechanical position of the stylus tip in relation to a reference point. Other systems use triangulation from sonic waves (G6-3D Sonic Digitizer) or light energy emitted from the stylus tip. Others (Polhemus) rely on magnetic field disturbances to locate the stylus tip within a limited work volume. Although manual digitizers have many applications and are relatively inexpensive and easy to use, they are not designed for acquiring a high-resolution surface mesh along complex curved surfaces. They also provide poor accuracy and resolution and are extremely labor intensive when high sample densities are required. They provide very little surface topology and therefore require great mathematical manipulation to reconstruct even the most rudimentary surface mesh.

Noncontact mechanical digitizers are emerging in the form of interactive microscopes (Reflex) and video data acquisition systems (VDAS). As visualization instruments, they offer wide opportunity for collecting measurement data but do not retain an available file of primary, topological data for import or export. Anthropologists and paleontologists can use the Reflex microscope to gather 3D coordinate data to compute measurements such as volumes and complex surface areas (MacLarnon, 1989; Teaford and Byrd, 1989).

It has been used to gather information on wear-related changes in tooth shape, angulation of wear facets, and area of dentine-free patches (Scott, 1981, 1985; Teaford and Oyen, 1989a-c; Teaford, 1990). Unlike many other systems, the Reflex system obtains three-dimensional measurements directly from the specimen (Fink, 1990).

The microscope works by carrying the object on the stage, which is fixed in the vertical plane but mobile in XY space. The measuring mark is moved only in the Z (vertical) direction. It gives repeatability spreads of less than 0.005 mm in X and Y and less than 0.02 mm in Z, according to the manufacturer (Scott, 1981, 1985), and in recent trials yielded standard error of mean ranges of 0.069-0.169 for objects at 5x magnification (Speculand et al., 1988), inter- and intra-observer measurements varying on 4.5 microns in the Y axis and 0.1 microns in the X axis (Adams et al., 1989), and root mean square errors of 0.005 mm in X and Y and 0.02 mm in Z (Setchell, 1984). Aside from the expected complicating factors such as object surface reflectivity (Stilwell and Heath, 1987), precision improves with experience (Speculand et al., 1988) and the overall performance is very strong (Adams and Constant, 1988; Adams and Wilding, 1988; van der Bijl et al., 1989; Hartwig, unpublished data).

The specimen size is limited to the working area of the microscope stage (110 x 110 x 125 mm). Reflex microscopes are well within the means of laboratory start-up costs but are not yet widely distributed in the United States, especially within research museums. They appear to offer a new level of comprehensive, three-dimensional, landmark-based data acquisition for user-specific statistical manipulation and analysis. As such, they are similar to the previously mentioned mechanical digitizers: necessary for improving the analytical foundation of traditional morphometrics but not designed for whole-object visualization or topological control.

Video data acquisition systems are practically limited to two-dimensional landmark and contour (boundary detection) analysis, but have a much wider acceptable specimen-size range, because the images are captured numerically as pixel signals. The same optical principles that enable digital image collection, however, also tend to distort the displayed image if the detail present in the original specimen is at a finer scale than the digital sampling interval (MacLeod, 1990). A VDAS can be relatively simple to install and operate as a stand-alone system, given the right combination of hardware and software components (Fink, 1990). Because the data collection device, strictly speaking, is a portable video camera, these systems are especially useful for archiving museum specimens while traveling for future two-dimensional morphometric analysis in the VDAS laboratory.

Photographic Methods

Passive digitizing techniques (i.e., feature extraction from images of subjects illuminated with nondirected lighting) have become popular due to the recent availability of digital cameras, frame grabbers, and related image-processing techniques. The methods attempt to mimic human vision. They provide proximity information and general shape information, but they are not appropriate for surface measurement purposes. Passive stereo methods (stereophotogrammetry), which extract range information by measuring the image shift of an object from two different perspective views, can provide relatively high accuracy, but they provide extremely poor resolution and reliability.

Automated Data Acquisition

Robotic Digitization

Of all available systems, tactile 3D robotic digitizers (machines that physically touch the surface of an object) provide unparalleled accuracy and resolution. This requires that precautions be built into the system to avoid destruction of the object. Obstruction avoidance becomes an almost prohibitive problem for complex objects because of the sophistication of the sensing system that is required. Robotic digitizers of this type are generally very expensive and require a great deal of a priori knowledge about the object prior to digitization. Application of this type of digitization system is primarily limited to industrial object verification applications where the objects are simple in shape and environmentally well understood.

Laser Scanning

Six degrees of spatial freedom (DOF) exist in nature: 3-DOFs for translation, 2-DOFs for rotation, and 1-DOF for scale. When using laser beams, four forms of relative motion between an object and a laser beam are needed to illuminate any point on an exposed surface of a three-dimensional object. Specifically, the digitizer needs two translational and two rotational DOFs between the ranging system and the object to provide full 3D sampling capabilities. The practicality of developing a 4-DOF digitizer is questionable due to mechanical, operational, and data-processing considerations. The fourth degree of freedom can be simulated from a 3-DOF scanner (just as the third DOF can be simulated on a 2-DOF scanner) by digitizing the object several times in different orientations. This approach requires that the multiple data sets be merged into a single-surface mesh. The shape of the object should dictate what type of scanning is used to best capture the surface information. Flat objects are best digitized using two translational DOFs. Convex objects require both a translation and a rotational DOF. Complex objects with non-axial geometry, multiple contours, and concavities (such as skulls or innominates) require a 3-DOF scanning device (see figure 4).

The most popular type of 3D digitizers utilize active triangulation of light. These devices represent one of the best approaches available for accurate, nondestructive, fast surface measurement. A single plane of light is projected onto the object and creates a linear light contour that is broken up and quickly digitized by the CCD (charge-coupled device) detector. This produces a large number of points for each line, typically equal to the resolution of the CCD array detector. The use of patterned light projection has increased sampling speed.

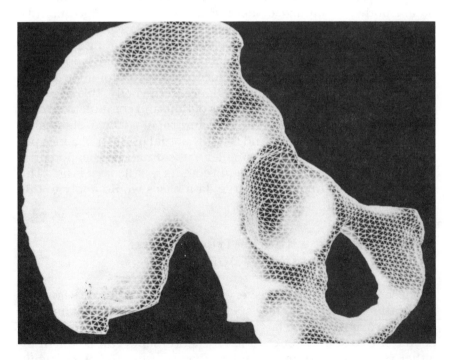

Figure 4. *This image of a human innominate was produced from data collected by a Digibot laser scanner. Objects with concavities, holes, and multiple axes pose particular problems to surface data scanning devices.* (Image courtesy of Steve Koch, Digibot Corporation.)

Sophisticated light patterns require specialized, two-dimensional CCD cameras, specialized facilities devoted to data collection, and extensive image processing hardware and software. As a result, these systems are generally expensive, complicated, unreliable, and limited in application. Other problems involve the limitations of the systems to 2-DOFs, which means that the subject needs to be composed of simple convex surfaces. Sampling for additional DOFs results in massively redundant data files that require identification and removal by computationally expensive processing to produce a final data set.

Several commercially available devices are hardware independent and as such are true computer peripherals. Others are commercial turn-key systems that are sold as complete packages for specific applications. Most of the devices could be classified as being moderately expensive, although the addition of a high-level computer platform can more than double the cost of a system. The advantage of these laser scanning devices is that they are semi-automatic and relatively simple to operate, are accurate (accuracy varies by type), and offer high resolution of organized topological data. Another advantage of laser scanning is that a large quantity of machine-readable data can be collected in a relatively short period of time; however, in the process of electronically curating a museum skeletal collection, for example, each individual bone would need to be done separately, which is not the case for volumetric (CT) scanners. With over two hundred bones in the body, this becomes a massive undertaking both in terms of time and computerized data storage. Even assuming that the anthropological skeletons in question are not complete, and that the actual number of bones per skeleton is a considerably smaller subsample of the potential number, the size of the data requirements is still very large. The problems of how to use very large data bases would apply to this approach as well.

Laser Scanning with One Translational DOF and One Rotational DOF

These 2-DOF scanners are very popular for a variety of scanning applications. Most systems, such as the Echo Laser Scanner by Cyberware, position the laser line to pass orthogonally through the center of rotation. A set of radial measurements can be taken by either rotating the ranging device or by rotating the object while translating the ranging system. This scanning system is most useful for convex objects because of the limitations imposed by surface normals that deviate significantly from the radial direction. The data grids produced by these devices are very attractive because they have complete topology, are very fast (the Echo Laser Scanner collects 250,000 data points in fifteen seconds), platform independent, inexpensive, and are easy to use (see figure 5, pages 206–7). Without the third DOF, digitizing complex objects (objects that have contours that are off-axis or have multiple axes, concave surfaces, or multiple contours) is difficult. As mentioned earlier, merging multiple scans can generate a pseudo 3-DOF (Sadler et al., in press), but limitations remain that prevent scanning of all possible object shapes.

Laser Scanning with 3-DOFs

The work volume of all 3-DOF scanners is cylindrical. The usual configuration includes one rotation of the object, and both a horizontal and vertical translation of the laser line. The Digibot Three Dimensional Digitizing System uses adaptive 3-DOF scanning by using the standard 2-DOF movement of the ranging system while rotating the object. This procedure is very useful for digitizing long, curved objects, such as a rib. Object rotations and horizontal translation of the ranging system along the object's surface while maintaining a minimal surface normal with the laser beam are very effective for gathering data on objects with deep concavities, such as orbits (see figure 6, pages 208–9).

Three DOF scanning offers the advantage of being more flexible with regard to the type of object to be scanned. It is automated, has efficient and complete geometry and topology combined in a single scan, is easy to use, platform independent, and relatively inexpensive. The additional degree of freedom requires mechanical movement of the object, which slows the data collection process considerably.

Dental Data Collection Devices

Mapping and visualizing the geometrically complex surfaces of teeth has a long history as a research problem in biostereometrics (Altschuler, 1976; Altschuler et al., 1981) and has obvious applications in physical anthropology. The age of mesio-distal and bucco-lingual distances as arbiters of taxonomy and adaptation is soon to be an historical footnote in light of the high-resolution visualization systems under development in the dental industry (see figure 7, page 210). As noted above, there is no a priori reason that physical anthropologists studying primate tooth morphology should not be coordinated with orthodontic and prosthetic dentistry laboratories.

Figure 5. *Image series of a live human face obtained from an Echo laser scanner demonstrating three modes of data display:* (below) *raw point data;* (page 207, top) *wireform display showing topological regularity of the database;* (page 207, bottom) *Goroud-shading of the polygonal mesh displayed in the photo above it.*

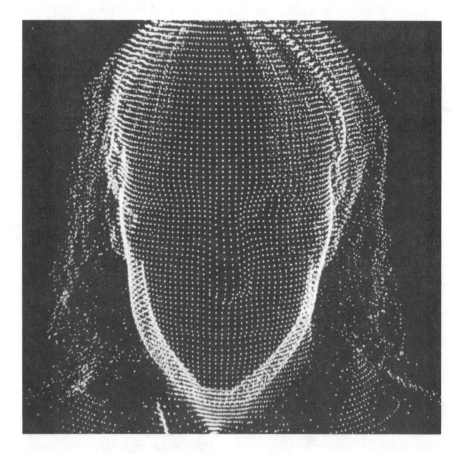

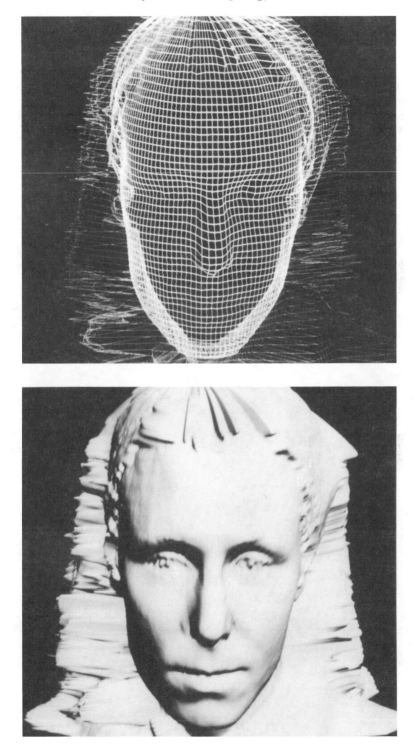

Figure 6. *A dry human skull* (below); *scanned by a 3-DOF laser digitizer is depicted in wiremesh* (page 209, top); *and Goroud-shaded formats* (page 209, bottom). *The figure on page 209 top demonstrates adaptive filtering (approx. 3,000 points) of the original data set (approx. 80,000 points) to reduce the quantity of data necessary for interactive display and rapid communication across computer networks.* (Courtesy of Colorado State University and Digibot Corporation.)

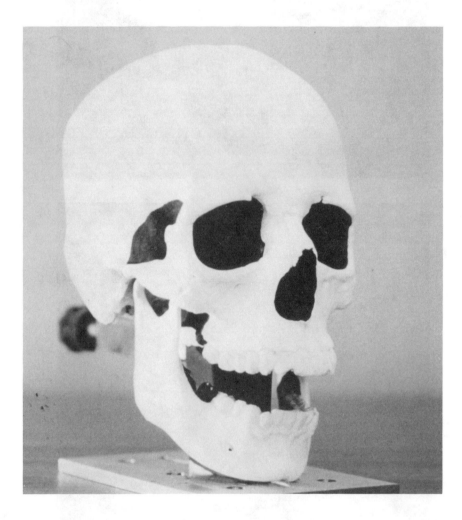

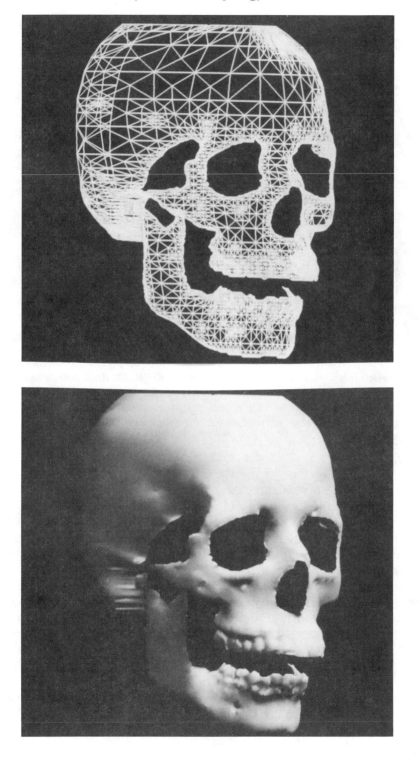

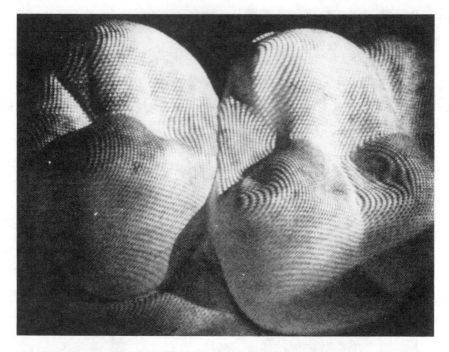

Figure 7. *Occlusal surfaces of human teeth imaged by laser light stereo-photogrammetry that accesses approximately 150,000 data points in less than one second. The technique requires multiple scans for complete (360 degrees) visualization of the object surface.* (Courtesy of Col. Bruce Altschuler, D.D.S., U.S. Army Dental Research Institute.)

Many of the same strategies for large specimen data collection have been applied to the problem of data collection at the small object level. These techniques involve the use of CAD/CAM technology to produce a computer-designed restoration with an accuracy of 25 to 50 microns (Leinfelder et al., 1989; Rekow, 1987; Duret et al., 1988; Bayne, 1989). Capturing the occlusal surface as a mini-landscape of three-dimensional data points invites a variety of topological, fractal, and experimental simulation studies. Several systems are currently under development, and the following are among those in use or in the trial stage.

Duret System

Developed by Francois Duret and Hennson International, this system uses a laser light imaging system to gather 3D data from the tooth. The system has been under development for over twenty years and is not yet in production, although several prototype systems are in use and have been shown at association meetings for the past several years (Duret et al., 1988).

Cerec System

The Cerec system has been in use in Europe for several years and has only recently been introduced in the United States. The data collection technique employed by Cerec involves an infrared light that is positioned over the tooth and reflects back to a lens and to a receiver where the intensity of the light is recorded. The surface of the tooth must be coated (usually TiO2 powder) to reflect uniformly.

Minnesota System

The Minnesota system, designed by Dianne Rekow, is undergoing field trials at this time. The collection system used to digitize the tooth is stereophotographic with multiple 35 mm photos being taken through a laryngoscope. Three-dimensional coordinates are derived from the stereophotographs (see figure 8). Kinematic analyses are conducted to reconstruct the path of the opposing dentition to better design tooth restoration (Rekow and Erdman, 1985; Rekow, 1987, 1989).

Figure 8. *Tooth row of the fossil primate genus Oreopithecus imaged from a dental stereophotogrammetric scanning system. The original data were translated using a custom-designed software interface and integrated into a CAD/CAM package for display and analysis.* (Courtesy of Dianne Rekow and Alfred L. Rosenberger.)

Software

Software for the sciences and particularly physical anthropology is a special problem, and one that is too large in scope to be covered in detail here. Software creation, particularly custom-designed software, usually lags behind hardware in development. Too often disciplinary scientists—physical anthropologists, biologists, anatomists, and so on—have been forced to divert from their primary scientific investigations to develop the software they need. To be fully equipped to tackle the major research directions in these scientific disciplines, part of a scientist's productive professional life is required to develop the appropriate software (perhaps reinventing programs that already exist elsewhere) rather than being devoted to the basic science, simply because the needed software is not available. From a commercial software development perspective, the situation is similar to that of "orphan drugs," since there is not a large enough market for vendors to be willing to develop what is required. Those scientists that have, through necessity, developed computer laboratories that support programmers in developing their specialized software must also develop a significant infrastructure in hardware and ancillary equipment that requires continuing research and university dollars. The collaboratory approach, particularly with the participation of corporate interests, should reduce significantly the need for in-house computing support staff, while at the same time increase the voice of basic science research in software development decisions.

Analysis (3D shape analysis, finite element scaling and analysis) and manufacturing strategies (for museum display and sharing physical specimens) require different formats for coupling with existing software programs. In addition, pointers in the data can be used to link the data to existing documents, photographs, maps, charts, diagrams, and other associated artifacts and materials, as long as the information is coded for the appropriate links. This hypermedia type of environment would be most useful for preserving the provenance of the data as well as supplying the appropriate data on previously completed analyses of the same specimens.

The Visualization Goal
A Digital Image Archive

In the following section we outline general specifications of an image storage and retrieval system appropriate to paleoanthropology. The components of such a system are already operational in existing information science contexts:

- acquisition devices in biomedicine
- automated information networks in academic library cataloging (and many other contexts)
- computer hardware and peripherals

Connections specific to the research designs of physical anthropology need to be made among these integral components. The initiative for creating, maintaining, and augmenting such a digital image archive must come from a consensus movement within the field and in collaboration with related movements in other academic disciplines and corporate-sponsored research.

An important early part of the data management cycle is to determine what data will be collected, what will be saved, what will be on-line, and what will be archived. It is not realistic to expect that every bit of data that is collected will be on-line at any one time. The amount of information that needs to be collected will produce literal data fire hydrants that would spew data sets beyond the requirements of any single investigator. A thorough understanding of the potential uses to which the data might be put should be undertaken prior to the development of any storage strategy. Data must undergo a cost-benefit analysis that takes into account the uses of the data sets.

Media

The vast amount of computer memory required to store visualization data requires careful selection of storage media. Ultimate storage should be on a medium that offers longevity and minimal space requirements. Storage media accessible to the individual user should offer rapid operating time. Decisions regarding media for importing and exporting of images are linked directly to the strategy of archiving them in a central bank, the structure and management of which are discussed in more detail below.

Hard Disk

The hard disk is usually the storage device within a single computer or computer system, although some hard disks are removable and therefore can be physically transferred from one machine to another. Disk fields are sometimes constructed by linking several hard disks together to increase the storage capacity of a system. Servers, minicomputers, and mainframes use this storage strategy. Clients can "hang off" of the system and use data that are anywhere in the system as though it was resident on the client machine. This is the terminal/mainframe, client/server model that is transparent to the user in a well-designed system.

Tape

Tape systems are often seen as "back-up" systems to protect valuable data from loss or damage. These tapes come in a variety of sizes and formats, from large open reels to small cassettes, which hold varying amounts of data. Many tape systems operate in conjunction with the operating system so that information on the tape can be accessed transparently from the workstation, thus expanding the storage capacity of the system. Usually tape access is significantly slower than on the hard drive, so it is somewhat less convenient. In addition, data stored on multiple tapes require physically changing tapes to get to those data. Nevertheless, tape offers an inexpensive, attractive, alternate solution to adding more hard disks. It is especially useful for storing data that are only required infrequently and need not be on-line at all times. A significant problem with magnetic tape is that certain environmental requirements are necessary to protect the tapes from damage. Even under the most controlled conditions, magnetic tape disintegrates in ten to fifteen years. Tape management requires rerecording at intervals of less than ten-year increments to protect the data.

Optical Disks (Platters)

Optical disk (read/write) storage is a relatively new medium that is still somewhat slower in response time and slightly more expensive than magnetic tape, but is a much more compact and archival storage medium. These platters come in several sizes and both sides of the disk can be used in recording. Optical recorders/players can be attached to servers, minicomputers, or mainframes to extend the storage capacity of the system. Optical "jukeboxes" operate like phonographic jukeboxes in that they have a mechanical device that searches for a given platter, retrieves it, and places it in the player. Response time is limited by the efficiency of the mechanical components housing the disks, rather than the disks themselves. Optical disk technology is ideal for storing primary data that may need to be accessed rarely, in an archival format requiring little or no maintenance.

Optical WORM (write once, read many times) disks are used in picture archive and communication systems (PACS) that were developed for radiological image data during the last decade and have since emerged as a very convenient means of storing large quantities of image data (Kenny, 1979; Imasato, 1985; Parkin et al., 1989, 1990; Brown and Krishnamurthy, 1990; Hofland et al., 1990; Juni, 1990; van Poppel et al., 1990). They are preferred over hardcopy x-ray film or magnetic tape, given their storage advantage (Kenny, 1979; Vannier and Conroy, 1989a), potentially faster access time (Parkin et al., 1990), and longevity. Magnetic tapes are written and read serially, thus slowing access time, and must be rewritten periodically to preserve image quality. Reels of magnetic tape are relatively expensive when their storage capacity and shelf-space requirements are considered (Jost and Mankovich, 1988; Parkin et al., 1990). The optical WORM system enables the user to control retrieval and analytical parameters for series of like images, and thus presents a baseline protocol for software custom fitted to research rather than diagnostic data.

PACS will be widely implemented only when they demonstrate a clear net profit to the purchaser or developer (van Poppel et al., 1990). Because of the initial high costs of current computing hardware, the point of break-even may seem too distant for host museums and universities with smaller operating budgets. This cost is compounded when existing acquisition devices must be married to new workstations. This retrospective approach entails the cost of interfacing two systems with widely different disk standards and formats. For this reason, development efforts have been directed at stand-alone image archive systems consisting of optical disk drives and personal computers (Juni, 1990; Parkin et al., 1990). There is little doubt that the cost-benefit curve for workstation performance will rise steadily in the years to come (Vannier and Conroy, 1989a; van Poppel et al., 1990).

Archiving Strategies

The movement to repatriate Native American skeletal and artifactual remains has gained momentum steadily for the last ten years, and most recently was brought closer to fulfillment by the passing of S.1980, a broad legislative measure requiring all federally funded institutions to transfer ownership of these resources. The language of this bill indicates that the rightful ownership and scientific value of museum collections are no longer at issue politically. Reburial will take place, sooner or later and to what extent depending upon litigation. The efforts of museum curators to indicate the research value of such collections and the need to permanently curate them (Ubelaker and Grant, 1989; Ubelaker, 1990a, b) have fallen far short of the political advantages of repatriation. The loss is similar to that a library would face if it had to destroy a certain percentage of its collection; how frequently the imperiled collection is used is no reflection of its scientific merit or future utility. Maintaining large collections of primary data is expensive, but not nearly as costly as losing a part of it. The reburial issue has precipitated action toward developing the archiving and curating strategies necessary to meet the problems of permanent data acquisition from temporary holdings. Visualization plays a critical role in offsetting the effects of repatriation by providing technical options for capturing as much primary data from the original physical specimen as possible. Curating those data in an image archive is the key to renovating the infrastructure of quantitative physical anthropology.

Structure

The structure of a visualization archive would enable the networked user to import selected images from a central bank, store them locally, and apply whatever analytical techniques are desired without affecting the "original" image, which is permanently curated in the central bank. Universal standards for data acquisition and formatting are essential for the maintenance and augmentation of an automated image library; only the networked satellite workstations are site specific and autonomous.[2] Both the structure and the management of such a resource must be approached with top-down rather than bottom-up planning. The cost-effectiveness and research potential of such a resource depends entirely upon developing a functional and efficient "house" before buying any "furniture." In other words, all parameters of the archive must be in place and standardized before any data acquisition begins. This eliminates incompatible formatting and, more importantly, avails the archive to virtually unlimited expansion. This latter condition is critical if skeletal and fossil material from around the world is to be imaged on location and then accessed by the archive through automated networks for later redistribution and site-specific manipulation (see figure 9).

The research advantages of an image archive are fundamental and represent, we believe, a quantum advance in basic science. No longer will an individual spend time and money visiting many different museums in order to measure enough specimens; the archive enables like collections from any museum to be collapsed into a single, cumulative collection available for export to the user (see figure 10). This at once enhances the available database while eliminating the redundant costs of traditional travel and per diem budgets. More importantly, handling of original specimens will be reduced to a minimum. Beyond its capacity to relieve the loss of information due to reburial, the image archive will become in the future a necessary database for the study of prehistory in the future.

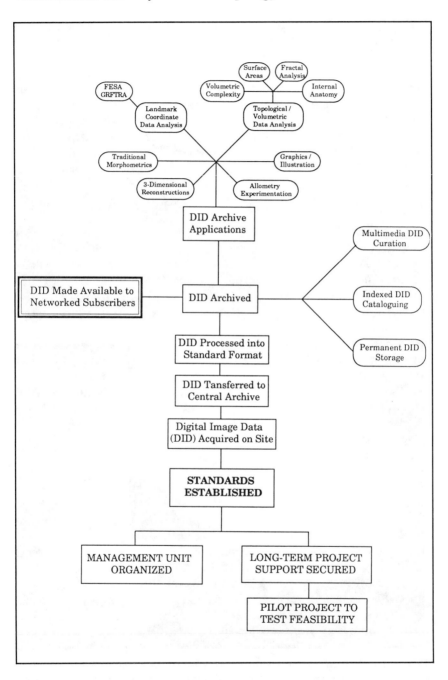

Figure 9. *Conceptual flowchart of a digital image archive from idea* (bottom) *to application* (top) *in physical anthropology* (see table 1 for definitions).

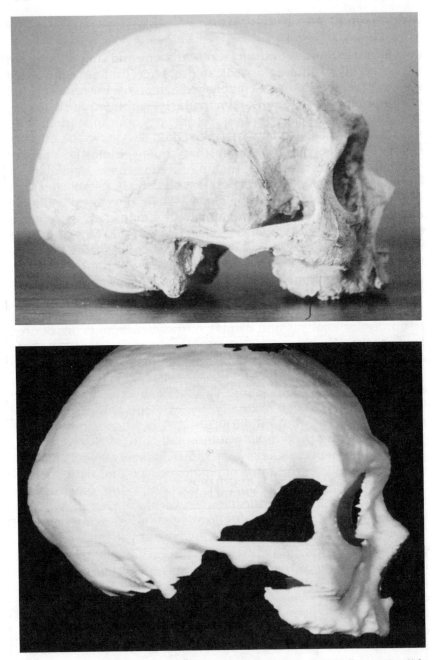

Figure 10. *Visualization technology can create a "global museum" by collecting image data on anthropological specimens such as this fossil hominid cranium (top) and by storing the images (bottom) in a digital archive accessible to any networked user. This image was produced from a Digibot 3-DOF laser digitizer.* (Courtesy of Colorado State University and Digibot Corporation.)

Management and Networking

Just as museums have curators, researchers, and support staff, an image archive requires a dedicated service unit. A centralized resource facility should be established to manage the research and development component, particularly application software associated with the storage, retrieval, communication, and utilization of the data. This resource could also serve the purpose of a central distribution library that could handle the massive requirements involved in making on-line data available. A consistent data management practice needs to be established that is realistically costed out on a life-cycle basis, developed using the most appropriate technology, competently managed, and given a relatively high priority by funding agencies so that an appropriate level of data management can be assured. These ideals are not likely to be met in the near future, and certainly will not be achieved within the traditional routes of physical anthropology project funding.

Associated with this is the need for low-cost supercomputing resources for analysis and display of these enormous distributed data bases. Exercises such as 3D shape comparisons are computationally very expensive; if populations of samples were to be compared, it is doubtful that the samples could be run at today's level of workstation computing. The use of supercomputing in the life sciences has been virtually nonexistent compared to the physical sciences, but initiatives such as those that can be envisioned using these data would rapidly alter that inequity.

Automated networks have been in use for several years and are most commonly in the form of catalogued textual data such as Medline. Other networks that are coming on-line (such as the National Research and Education Network) are part of what are becoming information superhighways for the exchange of data. Traffic on the current networks is growing at the rate of 20 to 40 percent each month, roughly doubling every five or six months. Because of the rapid advances in technology, funding needs to be available for translation and migration of data across systems. Some data, collected by federal agencies ten or more years ago, cannot be used because the data were collected on technology that is no longer available. In addition, digital data archived on magnetic media must be rerecorded on a ten to fifteen-year cycle. Effective digital data management must ensure that long-term data retention is built into an archive's resources.

Discussion

An image archive obviously will not supply all the types of raw data required by physical anthropologists, but it will go a long way toward: (a) expanding the possible types of research undertaken in comparative morphology; (b) substituting for imperiled or otherwise inaccessible specimens; and (c) modernizing analytical methodology. Justifications for networking the archive data include the following, some of which are discussed in further detail below:

- enhances accessible database by collapsing disparate collections into a single collection, a "museum without walls"
- saves substantial costs in grant proposal budgets by alleviating redundant data collection by individual scientists
- archive access cuts time and associated costs for travel to primary data storage sites and for the data collection phase of research
- democratization of data access across disciplines and between institutions
- data analyses can be conducted through networks on data and with machines that are not resident at the researcher's site, opening opportunities for collaboration and the foundation of "collabs"

The nature of research in paleoanthropology is such that the ratio of investigators to the "amount" of primary data is unusually high. In some specialties, such as human paleontology, the ratio at times approaches 1:1. Assembling the large databases necessary for statistically significant quantitative analysis is a particular logistical problem. For rare or underrepresented categories, an individual researcher may have to travel between or among continents to access enough specimens. This leads to reiterative travel and per diem costs, as well as substantial time lost in the data collection phase of research. For most morphometric analyses, handling of the original specimens is not necessary and often contributes to specimen deterioration. Several museum primate collections in the United States reflect the inevitable results of frequent use.

As part of an integrated image archive, on-line databases deliver a highly useful and versatile rendering of the original specimen to the networked user. The data, rather than the user, are mobile, thus eliminating redundant access costs. The user defines the search parameters for each import request from the archive in much the same way that on-line card catalogues are designed. Because the archive is assembled according to universal data acquisition standards, the remote user can efficiently "travel around the world" collecting specimen images without ever leaving the laboratory or spending grant dollars. More importantly for the long run, museum curators are able to reduce handling of original specimens to a minimum.

Concluding Thoughts

As presented here, visualization is a multidimensional and multidisciplinary tool for conducting basic science research. Aside from the practical advantages of automating a large segment of the empirical database, visualization opens new windows of investigation simply by virtue of computer graphics capability. The advantages of visualizing morphological data include the following:

- eliminates size handicaps: because data are stored and projected as voxels (volumetric pixels), the size of the displayed image can be manipulated without necessarily altering the specimen topology
- accepts and combines multiple modalities of data acquisition
- reduces the level of abstraction between morphology and analysis
- greatly enhances modelling and 3D reconstruction techniques, as well as illustration options: the inherent capacity of digital images to be enhanced, mirrored, and selectively distorted would offer researchers new options for experimenting with fragmentary fossil remains
- delivers immediate access to new analytical categories such as topologically intact surfaces, object volumes, pattern matching, and segmentation

As morphologists look ahead to shape analysis in the next century, automation is becoming imperative rather than selective. Achieving a balance between proximate expenses and ultimate benefits will make the difference between perpetuating outdated modes of research and capitalizing on a collective vision. The impact of sequential innovations in visualization technology on the field of physical anthropology has been substantial and is ongoing. The relatively recent emergence of information systems science and biomedical visualization promises to move much of basic science research to a new level of operation. Physical anthropology is in an opportune position to anchor collaborative exchanges among diverse academic and corporate interests, and to signal the potential of advanced data acquisition and management strategies.

In the space of little more than twenty years, computers in the academic environment have evolved from experimental and expensive to essential and cost-effective. The agenda proposed here for developing and implementing a visualization archive in anthropology and biomedicine has only begun to evolve. The archive as a functioning research tool is still years if not a generation away. The practical, theoretical, and analytical advantages to collecting and distributing data in this way are easy to justify but obviously very difficult to deliver. Only time will distinguish speculation from realization. If the latter prevails, the end result may be a methodological shift within physical anthropology that has the potential of centering the field within the activities of many progressive technological disciplines.

Endnotes

[1] The collaboratory concept was first proposed at an invitational workshop convened at Rockefeller University, March 17-18, 1989: "collaboratory," as the electronic analog of existing labs, a combination of the words laboratory and collaboration. Collabs contain the same resources, people, instruments, and databases as an independent laboratory, but the people and resources need not be in the same physical space; whether they are in the same room, state, or continent makes no difference. In an actual, physical laboratory papers are written, data collected and visualized, experiments performed, and analyses carried out; collabs are capable of all these activities, but at considerably less expense and at much higher capacity. Technology for the physical connections exists and is operational, but is only beginning to be exploited. The conventional approaches of small group interactions and small-scale interdisciplinary work at remote locations (most notably at museum venues, where unique resources are made available to the group) will soon be consumed by the advantages of electronically driven collaboratory communication.

[2] An analogous problem can be found in the in the exchange of data between scientists and other users in cartography. The Department of the Interior has established the Interior Digital Cartography Coordinating Committee and the Office of Management and the Budget has established the Federal Interagency Coordinating Committee on Digital Cartography, which are responsible for establishing standards, formatting data exchange, and maintaining facilities for exchange of LANDSAT data based in Rapid City, SD, through the mainframe in Reston, VA, over GEONET. Input digitization is carried out at regional locations. These data are used by a variety of different scientific disciplines, performing varied analyses, in a framework parallel to the one proposed here.

10

Evolution and Education

S. L. Washburn

This final chapter is a personal exposition of the multifaceted applications of evolutionary studies, as they could be applied to education. It follows Washburn's earlier publication (Daedalus, 1974).

The need for schools in a modern, technology-based society is so obvious that it is easy to forget that, through most of human history, people learned languages, complex social systems, and elaborate technology without formal schooling. In pre-agricultural societies, people learned the local flora and fauna, geography, economic practices, social customs, religion, and folklore — a whole complex way of life. Although the high school graduate has learned different things than has the gatherer-hunter, it is not at all certain that they have acquired as much useful information.

The modern student has the enormous handicap in that what he or she learns seldom fits into any readily visible way of life. In the folk society, learning is motivated by identification, emotion, and clear goals. All the information disseminated is readily related and applicable to living in that particular society; learning is not removed from the context of living. From an evolutionary point of view, these motivations provide the necessary conditions of learning. What may be learned depends on biology, but what actually *is* learned depends on the social system.

The act of throwing might usefully illustrate the point. People easily learn to enjoy throwing. Many games include throwing, and throwing, like all skills, requires an enormous amount of practice. From an evolutionary point of view, throwing has been an important skill to the survival of the species. It was important in hunting and in warfare, and practice in play ensured the development of the necessary skill. For example, to ensure the ability of being able to throw a harpoon effectively while sitting in a kayak requires that training start in childhood. This exceedingly difficult skill, which is later employed by adults in the most dangerous situations, requires years of practice. The child, throughout his or her upbringing, sees the practical importance of the skill and the social rewards received by the successful hunter. Thus, the child's identification with adult success guarantees the pleasure of play and practice.

Chimpanzees are one of the very few animals that also throw. They have the anatomy that permits throwing either underhand or overhand (brachiation), an anatomy shared only with the other apes and humans. But they do not practice throwing: a chimpanzee does not make a pile of rocks and throw them at a selected target, and other chimpanzees do not watch and call encouragement. In humans, however, the act must be embedded in a social system of sports, so that the years of repetition needed to develop a high level of skill are fun. That is, the pleasure of play is the biological solution to the motivation for learning. The same point could be made about other aspects of adult life. Social relations, technical skills, art, and dance—all visible activities—are practiced in play years long before they are used in adult life.

The learning of skills has been so important in human evolution that natural selection has incorporated the basis for learning skills into the structure of the brain. The part of the human brain most closely related to hand movements is very large, and proportionately far larger in human beings than in any other primate. But the fact that a biological basis exists for learning skills does not mean that the skills will be learned. Learning occurs only within a social system that puts the learning into an acceptable form and that clearly shows the learner the utility of the skill in adult life. In a folk society dependent on the spear in both hunting and war, the games of childhood are direct preparation for clearly perceived and valuable adult actions.

Learning of skills is affected by rewards, but the nature of the animal greatly affects the kind of reward it appreciates. In the matter of hunting and fishing, human beings, especially males, will spend many hours and make great efforts for the most minimal success. For many people, catching a fish is a reward far beyond the possible economic value of the food. Many states have departments of fish and game that began spending millions of dollars to provide fish and game for hunters long before programs for hungry children were even considered. Clearly, there is something about hunting that is remarkably rewarding to human beings. The importance of hunting in our evolutionary history has built a human biology that makes hunting easy to learn, with very little compensation required to motivate very substantial efforts. The pleasure of hunting makes people willing, for a minimal reward, to work hard, practice the necessary skills, and devote considerable time and expense.

The education of people in gathering and hunting societies is successful without schools because learning is motivated by identification, emotion, and clear goals and because evolution (through natural selection) has produced a species for whom classes of learning are natural. Implicit in this discussion is a distinction between "learning" and "schools" which is most clearly illustrated in the first few years of life. In monkeys, observational and experimental studies show the importance of early learning. In a natural setting, the mother and infant are together constantly. The infant learns from its mother or from other animals in close proximity. Long before the infant monkey is weaned, for example, it has tried the kinds of foods its mother is eating, often spitting some of them out in distaste. In time the young monkey learns what is edible and what is not. These animal studies suggest the great importance of early experience, strong social bonds, and an enriched environment for learning. The emotional situation is a critical factor in learning; an emotionally deprived infant ceases to learn. If we turn to humans with these studies in mind, what they suggest is that learning in the first years of human life should be examined with the greatest care.

The schools receive the products of these critical early years, and it is an illusion to suppose that a single teacher in a large class can overcome the deficits of each child's early experience. In both monkeys and humans, early learning takes place in an environment of few individuals and strong emotions. It takes place over many hours each day, and there is enormous pleasurable repetition. A few hours a day in a large, well-disciplined class, without strong emotional bonds, cannot possibly provide the biosocial setting for effective early learning.

As a monkey grows older, much of its time is spent in a play group. Learning from peers gradually replaces learning from its mother, although the attachment to its mother may last for years after weaning. In the play group, the monkey learns those social skills which form part of later adult behavior, and these are practiced every day. At a later stage, juvenile females play with infants and become skilled mothers before the birth of their own infants. With time, the play between juvenile males becomes rough, and they practice the aggressive behaviors that will be essential in later life. Both sexes learn the behaviors that are essential for life in the social group, and these behaviors may differ from one group to the next, depending on the kinds of food and the specific dangers caused by predators.

From an evolutionary point of view, much of the behavior of human children is similar to that of other primates, but the time of maturation, learning, and practice is greatly prolonged. The slowing of maturation is costly from a biological point of view. Young primates are likely to be injured or killed, which suggests that prolonging youth must have been so important that it was favored by selection in spite of the risks involved. In the human, the delay of maturation took place long before the advent of agriculture, so the explanation of the delay must lie in the life of the hunter-gatherers. This biological delay appears to be directly related to the evolution of a nervous system that learns the technical and social complexities of the uniquely human way of life. Both the brain and the way of life evolved in a feedback relationship with each other.

We must also recognize that the problems of motivation for learning change with age. Humans are the slowest of the primates to develop, and human children mature at very different rates. Every teacher knows the great changes that take place with maturation, and yet the implications of maturation receive scant attention in planning the school agendas. For example, at ten or eleven years of age, girls spell better than boys. And at this age, the girls are nearly two years more mature than the boys. If anyone stated that eleven-year-old boys spelled better than nine and a half year olds, everyone would laugh. There are the strongest biological reasons for not putting children of the same chronological age into the same class. Learning and maturation are in a feedback relation; neither one proceeds without the other — but our culture stresses learning almost to the exclusion of maturation.

The need for schools and the problems of the schools come from three quite different sources. First, there is no longer any folk society, no highly visible system of human behavior that can be easily appreciated by all the participants. Second, many of the skills needed in modern complex society take years to acquire, and so far no one has succeeded in making the intermediate goals — for example, the steps between elementary science and being a chemist — exciting and adequately rewarding. Third, the necessary ways of life in complex society no longer bear any simple relation to human nature. In all folk societies, for example, basic learning is complete by maturity, and shortly thereafter, people live as adults. Today, particularly in training for professions, a large percentage of students have been mature for many years, and yet the form of education they receive is the same as it was in the elementary school years, a situation that is both unnecessary and a cause of constant friction between student and teacher. The institutional framework of the graduate school is merely an extension of the elementary school; there are still courses and marks, teachers and children. To live as an adult, a person must drop out of school. Many students are "preparing" for life for twenty-five to thirty years—a third of a lifetime—and that is clearly contrary to basic biology. In many other cultures, people are treated as full adults some two to four years after biological maturity.

Perhaps the most fundamental difference between folk learning and modern education is that, under folk conditions, most people did the same things. There was a division of labor between men and women, but aside from that, there were not many different kinds of jobs. Boys played with spears because all men would use spears in hunting and warfare. In marked contrast, very few students who take biology today become biologists. Most students realize that they will not directly use much of the information of their specific training. Lacking the initial desire to learn what will be useful, students become bored, and the school resorts to discipline to maintain the process of education. But discipline does not facilitate internal motivation, and examinations are no substitute for life. A species that tries to substitute discipline for pleasure in learning is contradicting its whole biological heritage. Human beings are not pigeons who may be taught to peck out the solutions to routine problems. They are the most creative, imaginative, social, and empathetic beings that exist. But schools may reduce youth to bored and alienated primates, people who have been educated out of their natural desires to learn while being separated from the larger society in which they must ultimately live.

Primates respond to encouragement and positive rewards, and human beings are no exception to this rule. Yet the schools discourage many students year after year with low marks and disparagement. It is contrary to human nature to achieve under such circumstances; thus, the customs of the schools may be one of the major reasons for academic failure. After enough discouragement, further effort does not seem worthwhile. The marking system in many schools rewards the few and discourages the many. Biology, I believe, correctly suggests that, after consideration of maturation (readiness to learn) and the social situation, the curriculum should be planned so that success (encouragement) is the norm. After all, the purpose of education is to help people to live happy, useful lives. Therefore, the educational customs that govern a third of a lifetime should be directed to those ultimate goals, and not to providing failure, discouragement, or exclusion.

To summarize what I have been saying, the study of animal

behavior and evolution suggests that early learning comes primarily from the mother, closely associated people, or parental substitutes (as in the kibbutz). It involves countless repetitions in a warm, emotional environment. This early learning is critical in preparing the child for learning in school. In our society, there is no substitute for parents talking, reading, and playing with their children. As the child grows older, peers become more and more important, but the home attitudes and actions remain of great importance. Children play with peers, and this is a fundamental setting for learning, but their games reflect the actions and values of their adult community. It is only recently, for special reasons, that a separate institution (the school) has become necessary for teaching the young, and its problems arise from the loss of the traditional folk learning situation in which learning depends on identification, emotion, and clearly visible goals.

From a biological point of view, perhaps the most fundamental change that could be made in the schools is to have peers as teachers. From a practical point of view, this would mean that slightly older children would do some of the teaching of the younger children. This could produce three important results. First, it would increase the number of teachers, so that much of the teaching could be done in the small, informal, emotionally supportive groups that are natural for humans. Second, it would provide satisfactory social roles for many children, roles in which they could be proud of their accomplishments and work for which they could be praised. Third, to teach successfully, peer teachers would have to really learn the subject matter.

Peer teaching has been tried with college seniors teaching freshmen, and there is no doubt that the quality of college education could be greatly increased tomorrow by the careful use of student teachers. But the essential point is that the slightly older peer can have a very different relationship with the pupil. The point of teaching by peers is to restore the conditions of learning that are natural in the kind of situations in which the human brain evolved. Just as many sports perpetuate natural situations of maturing and learning—situations that are pleasurable and socially rewarding even if one has to practice long and hard—so peer teaching could be used to create intellectual situations that would be pleasurable and rewarding even if one's intellect had to practice long and hard.

Peer teaching provides two kinds of motivations. The learner sees that someone just slightly older can do the required tasks and, as a result, no longer considers the tasks as impossible or irrelevant. Another motivation is seeing the peer teacher have an important social position in the life of the school, just as the successful athlete does. A peer teacher might spend up to one-fifth of his or her school time helping peers. Given the many different subject matters, most children might be teachers at least occasionally, and all would have the opportunity to be helped by peers. In the past, without schools, human beings were able to master complex social tasks in ways that were in accord with human nature. With modern knowledge, science and a whole profession of educators, surely we should not do worse.

The Need for Clearly Apparent Goals

Even if early learning is successful, and even if peers help in creating a social situation for elementary learning, the problem of goals remains. How can the years in school be made to seem important to the developing human being? The answer is that they have to be important in ways that a young person can understand. Many children do not enjoy learning simply because they do not believe that what they are required to learn is practical and of any importance. Many older people are skeptical of what is required because they too doubt that what is taught in the schools — beyond the basic skills — is necessary. Take Latin, for example. For many years, Latin was taught because it was the language of scholarship, international learning, and religion. But when it ceased to have these functions, it was still taught for "mind training." Latin is essential if one is to understand the classics and much of European history, and probably too few people master it today. But it is a lasting blot on the history of education that millions of hours have been spent to learn a language that is, for most people, useless. The learning can only be justified for spurious reasons.

The motivation of the older student requires support and understanding in the home and in the community. The student needs visible, important goals, as in the folk society. As long as the school is considered a separate institution, insisting on tasks that bear no relation to life, it will fail in its most important intellectual functions. Since the motivation for learning comes from family, peers, and society, the isolated school has minimized its chances to teach.

The ultimate aims of education should be to give the student some understanding of human nature and the world we live in, intellectual fun, and preparation for jobs. These matters have become technical problems, and so we need to have an institution designed to deal with them.

The simple folk explanations of human nature are, however, no longer enough. We are living in the world of DNA, modern biology, and complicated medicine of incredible possibility. The small, flat world, the center of the universe, has been replaced by a universe of temporal and spatial dimensions that would have been totally incomprehensible to our ancestors. Music, art, and literature are available in forms and quantities that the kings of years ago could not have commanded. Even as late as 1900, most of the jobs in the United States were on farms, whereas the sheer variety of careers now open to young people is something entirely new in human history.

If these technical problems were simple ones, they could be managed in the home, just as learning was for countless thousands of years. They are no longer simple, however, and they are changing all the time. But human beings are not changing. If we are still evolving, the rate is too slow to be of practical importance. The biological nature of humans that evolved under the conditions of gathering and hunting is the nature with which we must learn to live in the modern technical world. This is a new problem. On the one hand, the system of education must consider the nature of the youth being educated, and it must consider the social, emotional, and biological conditions of the system. On the other hand, the schools must also show how to deal with the new problems arising from the rapid change in the technical sciences.

The test for the schools comes in the actual behavior of their graduates. How do they approach problems? What resources do they bring? Are they innovative, cooperative? Do they know where to look for knowledge and how to ask for advice? It is not easy to test for these important abilities, and again, biology suggests some of the complications. For example, those abilities that are measured by IQ tests are not changed by massive damage to the frontal lobes of the brain, a fact that was used as a defense of frontal brain operations. Yet this part of the brain is composed of billions of cells, both the number and complexity of which have increased in primate evolution, and especially in the later phases of human evolution. Since such increase is due to natural selection, we know that the frontal region must be important. Its probable functions are foresight, insight, planning, persistence, and originality, all of which are thought of as the higher mental functions. The so-called IQ of a child is greatly affected by their environment, but the IQ test does not measure these most important human abilities. Tests of performance clearly show the disastrous results of frontal brain operations, and the implications of this fact for testing are clear. Human beings are much too complicated to be evaluated by pencil-and-paper tests, which can be quickly administered to groups.

What counts in life is performance, and performance is based on both biology and experience. For example, in modern science most projects are cooperative and, in real life, the ability to work creatively with others is of the utmost importance. Yet the educational system minimizes cooperation and requires the student to work in isolation. The desire to grade a paper (how will we know who did the work?) forces immediate and trivial emphasis on individual work, which, in effect, discourages learning the complex personal and intellectual complications of cooperation. Yet performance in cooperative projects that lead toward problem solving is probably a far more useful measure of ability than anything measured by conventional tests.

The problem that arises from grading the individual may be illustrated by writing and spelling. If the goal is the ability to express oneself in useful ways, the student needs to be encouraged to engage in the repetition necessary to master expressive skills. But if the student is penalized too much for misspelling, the whole task becomes distasteful. This is particularly the case if difficult words are introduced merely to find out who are the best spellers. Animal behavior clearly shows that creatures who are repeatedly discouraged give up a task. Conversely, success leads to repeated performance and to the development of skills. If the classroom, through frequent testing and grading, becomes the symbol of discouragement, students can protect themselves only by withdrawal in one form or another. From a biological point of view, the classroom should be a place of encouragement, and students should run into class as joyfully as they run out into the play yard.

Many educators often talk about the importance of discipline and the learning of a foreign language. As noted earlier, mastery of a second language was not a problem when there was no school and the child was in a situation in which learning two languages was useful. Years of study in school may accomplish much less. Perhaps a minimum standard for evaluating a school might be that it not do worse than what was easily achieved without a school. Certainly, experience shows that learning a language is easy if the process begins early, if it is done with a native speaker, and if the child perceives it to be useful. A child may have a good start on a language but then forget it if he feels there is no reason to continue. Why it is useful to learn a language will depend on the time and the place. For example, in California today, there are millions of Spanish-speaking people. This is the foreign language most likely to be useful to citizens of that state. Accordingly, many California schools should offer Spanish, not for magic or mind training, but because it is useful. Spanish-speaking children could be the peer teachers, and many hours could be devoted to games, texts, and tapes. This would change the position of the Spanish-speaking child in the school, and it might even prepare the way for better understanding between the cultures. Later, Hispanic history could be studied, and the problems of Latin America might be contrasted with those of the United States. Finally, every effort could be made to show all the different kinds of careers in which it is an advantage to speak both Spanish and American English. Every aspect of a program aimed toward useful mastery of the language would be different from programs for learning a soon-to-be-forgotten smattering of a scholarly language.

What Value the Past?

In spite of our interest in the past, the slogan "back to basics" and the narrow and fallacious view of education that it represents is deeply disturbing. Clearly, many people believe that a time existed when tough-minded teachers really taught the fundamentals and that recent attempts at educational reform have been disastrous mistakes.

Perhaps at this point I should note that there are two very different uses of the past: we may study history in a search for fundamentals, for basics, and for customs that have survived because of their validity, or we may study history to see the endless flow of unnecessary error. We may view the past as what happened before modern science, technology, and education opened the way to a fundamentally new world, or we may regard it as the embodiment of customs that were right and will continue to be right.

Although our law allows for some change, it is basically conservative: it attempts to devise theories of what is right today from the customs of the past. This might be well and good except for the fact that many of those past customs were based on ignorance, and consequently, we should give them no more attention than we do the notion that the earth is flat. The Supreme Court's decision on the use of physical discipline is an excellent, though unfortunate, case in point. The philosophy behind it is that of an era when beating children both at home and in the school was considered normal and right. The underlying assumption being that discipline is the best way to motivate and influence behavior. Even more important, it presupposed that discipline is the key to education, that if students are not achieving we must give them more homework and return to rote recitation. These methods have a certain natural appeal to some educators. Reliance on discipline frees the teacher from the obligation to make subject matter meaningful and to provide intellectual motivation. Such methods may also raise test scores, no matter that such methods destroy the potential of the educational process, often the potential of the one being educated, and that they belong to an era that has passed.

We can understand the issue of motivation more clearly if we contrast the average student's behavior on the athletic field to his or her performance in the classroom. In the game, the goal is understood and socially accepted, and it appears important to the student. Consequently, the student accepts a great deal of practice and discipline as normal and necessary. In the classroom, however, a comparable exciting goal rarely motivates behavior.

The need for discipline, or even punishment, arises when the student does not understand the purposes of education. Before discussing the reasons why that is the case, however, we should keep in mind that the purposes of education and their relation to the larger society, of which the schools are only a part, have radically changed in the last few years.

A century ago, the three R's were considered an adequate education for all but an elite few. Higher education was considered necessary for only a few professionals, primarily the clergy and teachers. For them, classics and foreign languages comprised most of the curriculum.

Today, most jobs require advanced training of some sort. Consider the difference between the small farm of a century ago and the giant, mechanized farm of today, which is probably run by a college graduate. Not only has some post-secondary education become increasingly necessary for many kinds of jobs, but modern technical society in general requires an ever-increasing number of professionals, paraprofessionals, and people who are educated to cope with the world.

Education, far from being only for the elite, is the essential foundation of modern society. It is this ultimate, profound importance that gives modern education an almost entirely new role. In a modern democracy, people are required to make a large number of choices; they are asked not only to make decisions on political and social issues, but to assess the technical judgments about the effects of smoking, saccharin, or laetrile. The progress of technology, science, and knowledge that immediately affects human lives and the social system forces a new view of education. Certainly, there was never a time when the "back-to-basics" was less appropriate in the vocabulary of an educator.

The "basic" issue comprises two radically different views of human nature. First, according to the traditional or back-to-basics view, most students will not learn unless forced to do so and until a wide variety of educational practices provide the force. Second, according to the opposing view, human beings are active, creative, and eager to learn until the desire to learn has been extinguished by the schools.

If the second view seems extreme, consider that, throughout

almost all of human evolution, people learned complex ways of life without benefit of any formal education. They learned complex social systems, technologies, arts, and religions. If needed, they assimilated a second or third language. Children learned by seeing the behavior of adults and playing at those behaviors. The diversity and complexity of the behaviors learned in this way are the strongest argument and the best evidence that human beings want to learn and that they are supremely creative. And the ease with which they learn is the most distinctive characteristic of our species.

Judging from the fossil record, the increase in the size of the brain, which is partially responsible for the ease of learning (there was reorganization in the structure of the brain too), has taken place over the last three million years. Previously, the brain of our ancestors was no larger than the brain of contemporary apes. Again, judging from the fossil record, the evolution of the brain was complete some thirty thousand years ago, and since then the varieties of the human languages and other behaviors are the result of learning. Without schools, humans developed agriculture and civilization, and until recently societies managed with very few literate people.

To put matters in perspective, our ancestors lived as hunters and gatherers for many hundreds of thousands of years. During that period, our biology evolved to its present form; agriculture developed and spread over most of the world in less than ten thousand years; modern science evolved in less than two hundred years; and now new information and technology spread around the world almost instantly. The world that supported some fifteen million people now supports over four billion, and the existence and fate of such numbers depend on modern education. Of course, there is nothing new in this point of view, every underdeveloped country is busily adopting educational systems. What is new, however, is the insistence that these systems, largely borrowed from Europe, long antedate the problems that they are supposed to solve.

We need not suppose that our education system is reasonable or inevitable its present form. However, our common sense, which is embodied in custom and law, tells us that a system involving so many people with such important results, one that is well known to all of us, must somehow be correct or the best possible. More fundamentally, there is no reason to suppose that the basic social structure of our schools, their assumptions, or the way in which they are implemented, have any more validity other than that they are ours. As *our customs*, they will inevitably be protected, honored, studied, and improved a bit here and there, even though the study of the customs of other people, of primate behavior, and of basic biology all point to the need for radical change.

The life of a child in a modern school is separated from the life of the adult community. The child is expected to understand that what is being required in school is reasonable, even though the need to use the information may be delayed for years. This is why punishment and discipline have become substitutes for play, identification, and immediately perceived reality. Several consequences stem from this evolutionally new situation. The child loses the whole natural drive to learn; the juvenile frequently detects that the requirements are not really necessary but are merely imposed by the traditions of the schools; and, as the years pass, an increasing number of students believe that dropping out of school is the only way to start living an emotionally satisfying life.

In order to turn these negative consequences around and encourage the basic human motivations of learning and creativity, schools need to stress the importance and meaning of what is being learned. A close look at medical education should resolve any doubts one may have about what human beings will endure if the importance of the goal is clear. Medical students manage years of the most exacting kind of preparation, years of often bitter competition, because of the ultimate goal. The same point applies to other kinds of advanced education in which the goals are clear. The problems occur in the years before the ultimate goals are sharply defined, these years are particularly trying for students who come from backgrounds that do not strongly support the academic effort.

If the larger community clearly perceives and supports the goals of education, then students will develop greater personal motivation, which will lead to stronger academic efforts, and then discipline will become largely unnecessary. To see adult life around us in order to be able to prepare for it is the heritage of our species. Today's youth, however, has a double problem: the varieties of adult careers and lifestyles have greatly multiplied, and preparation for them has become enormously complex. It is not possible to return to the simpler life of our ancestors. But to regain the role of play, identification, and motivation, as it is built into our biology by hundreds of thousands of years of human success, teachers, parents, and peers must all help the student to see how the daily activity in the school is related to the important problems of living in a rapidly changing, technical world. The customs of our schools would look very different if we came to them with a background in animal behavior and with knowledge about the traditions of other peoples.

To summarize, the study of the customs of other peoples shows that human beings *want* to learn, but that they learn best when highly motivated by clearly defined, useful objectives. If we accept this premise, then the need for some changes in the present school system should be apparent. No one seriously believes that educational reform can be achieved quickly, but it is now time at least to begin.

Among other possible innovations, learning from peers as we have described it could be used as a way of revolutionizing the social structure of the schools. We know that it is effective because it is the major way in which human beings and other primates learn. It is not new; there has been both research and well-documented literature on the subject. It is a supremely important mode of learning that continues to be eliminated by adhering to the traditions of our schools and by returning to still more antiquated and fallacious concepts of how humans learn.

The "back-to-basics" movement, unfortunately, will leave children ill-prepared to deal with the complexities of our current world, not to mention a future that will hardly be less complex. Because its underlying philosophy runs counter to all we know about how children learn, it may, ironically enough, lead to the very thing its supporters are trying to remedy: *lower* academic achievement, as well as to narrow and out-of-date curricula and teaching practices that rely on coercion and repression.

Higher Education

Education should provide an understanding of humankind and the world, as well as offer vocational training. The process should be pleasant and continue throughout life. However, the traditional structure of the university based on departments prevents the attainment of these goals. If a student wants to understand the nature of the world, he will find the information scattered in departments of astronomy, chemistry, geology, mathematics, and physics. To take all the courses would be difficult enough, but even if the student had the prerequisites and the time, in the end he would have to make a synthesis which his professors have not made. Avenues of approach to the understanding of humans are far more labyrinthine. Numerous biological sciences, the social sciences, history, and the arts are relevant. In short, the university is divided by tradition, research, and what appears convenient for its faculty. Its divisions not only do not help the student attain broad intellectual goals, but they make it difficult for him to master knowledge from more than three or four closely related departments.

Many of the problems facing the world today do not fit into the traditional departments of educational categories. Population, energy, and crime, for example, involve many different disciplines. Each has a history and each involves trends. The daily newspaper provides numerous examples of problems the solution to which will not come from any one branch of knowledge. The educated citizen in a democracy today should be able to see what kinds of information are actually useful in moving toward better solutions for the real problems that face us all. Highly specialized education fails to prepare one for processing knowledge that is essential for democracy, for informed decision making. For example, the theory of continental drift provides understanding of earthquakes, mountains, volcanoes, and distributions of ancient life.

Clearly, no solution to the present educational dilemma exists. The task, therefore, is to start working toward different methods of education that will give the tools of knowledge to any intelligent person. Another solution is to supplement existing educational opportunities with new kinds of educational experience. This can be done by building courses around case studies that illuminate the kinds of contributions the sciences and humanities can make, rather than building courses on the survey method alone. The case-study method would present a proven technique that has been applied to problems of scientific and humanistic knowledge. Cases could be chosen to reveal the forefront of advancing knowledge on which progress rests and to point out the vocational possibilities. The cases studied should give a clear understanding of the changing nature of the scientific enterprise. They should show the fascination of knowledge and the power of techniques. They should help people to see the nature of a wide variety of sciences and to understand why highly specialized technical training is necessary if one is to contribute to the progress of any of the sciences. In addition to offering some understanding of the nature of the world and of science, the cases should show the diverse careers open to those trained in science. For example, the phenomena of continental drift creates a new perception of the earth's surface geology, its interior, and the process of mountain formation, as well as the background for biological evolution. Proof of drift involves obtaining cores, radiometric dating, and magnetic reversals and reveals the relations of techniques to the progress of science. The examination of ten to twenty such cases over the course of a year could give the student a perspective on the physical sciences: what they can and what they cannot do. By selecting cases that are both fundamental and controversial (such as continental drift, the origin of life, the composition of the moon), teachers could present science as an ongoing human activity. Likewise, the astronomy of quasars and black holes is radically different from that gleaned from the conventional telescope of not so many years ago.

The purpose of a case study approach is to give insight into the meaning of the various sciences. This approach should be contrasted with that set out in the elementary course, which is the first stage in training a professional. For example, a widely used elementary paleontology text contains some five thousand technical words and names. A major part of the effort in using such a text is the mastery of a vocabulary useful only if the student continues in that subject. The same text gives no sense of the excitement of the discovery of the fossils. The information is unrelated to either intellectual history or technical progress. One would not learn in its pages that the history of paleontology was closely related to anatomy, embryology, and genetics and that evolution can be studied as it is reflected in the structure of macromolecules. Case studies, in contrast, might stress the meaning of evolution, experimental analysis of the processes involved, and the nature of progress.

Immense progress has been made over the last century in the understanding of the physical world. The progress is clearly based on the development of methods which provide answers to basic questions. There has been no comparable progress in understanding the nature of man, and it is not even clear that people know how to ask useful questions. For example, it is often claimed that a knowledge of history is essential if we are to avoid the mistakes of the past. But is there really any evidence that historians understand the present in more useful ways than other people? Since a student cannot possibly take courses in the history of all peoples, the useful meaning of history requires some kind of abstraction, some generalizations from the boundless but uncertain "facts." In contrast to a course outlining the history of some country, a case study approach would present some body of historical information to show how this helps us to live today, to make more useful decisions. Case studies would not replace the traditional courses, but would call attention to the resources of history and would make it clear how these could be mined. If the nature of humankind is to be understood, there need be analyses of cases which show just how the past illuminates the present, just how history is useful. It is by no means obvious that the rapidly changing modern world can be operated effectively using the customs (laws, economics, political systems) of many years ago.

Starting with an effort to understand our species by a case study on the brain is one way to avoid the cultural biases that dominate the whole academic system. For example, warfare, physical violence, and torture have been an important part of human behavior, yet these receive scant attention in the curriculum. It has been estimated that in primitive fighting about 25 percent of males were killed. A case study on the relation of the brain to violence might open this area to discussion and show some of the fundamental relations of biology and behavior. Through his evolution, man has evolved a biology which easily allows him to learn aggressive behaviors. Biology does not make the behavior inevitable, but it makes it highly probable, unless controlled by custom. A case study on violence would demonstrate how biology, early experience, and custom interrelate to affect the frequency and nature of this kind of behavior. The study of aggressive behaviors is particularly useful in showing the problems which come from language. Words describing aggressive behaviors usually have highly emotional connotations, and they demonstrate how difficult it is to speak of the behaviors without using a vocabulary which implies the conclusions even before the events are fully described.

Behind the suggestion of a case study approach to understanding the nature of the world and humankind is the assumption that survey courses give only a superficial statement of the conclusions of a traditional piece of knowledge. In contrast, the cases would delve deeply into an important problem and show how it is related to technical information, history, literature. The essence of the well-chosen case would be in its ramifications.

Knowledge is changing rapidly and in important ways; to participate intelligently in modern democracy, citizens must be able to recognize the implications of an event or discovery and make choices in a way which is not possible at present. Continuing education is not a luxury, but an essential part of a modern democracy. Just as traditional classical education fit a way of life and sprang from a set of beliefs and values, so modern education must fit a new way of life. The primary contrast between the two is that the traditional system saw wisdom coming from the past and was built around the power of the Word. Present education must open the way to constant adjustment and change. Ancient texts no longer offer clues to reality.

The highly adaptable, emotional, social creature we call "human" learns best in the early years through emotionally rewarding interpersonal bonds and later through clearly visible, socially important objectives. In education, discipline is no substitute for internal motivation. In the young, play is the primary internal drive for learning. The repetition which is necessary to master skills is guaranteed by play in all cultures not dominated by European educational traditions. Our system separates education from life and substitutes examinations for experience. The classroom reduces the most intelligent, imaginative, exploratory primate to a bored robot, learning "facts," most of which form no part of a preparation for life. What we know about primate nature indicates that the present form of the educational system is not necessarily efficient or the most suited to our nature. Education should be aimed toward clearly defined goals and operated in accord with an understanding of human nature.

Many people now spend one-third of their lives getting educated. Surely this vast expenditure of time and human effort should be pleasurable, meaningful, and liberating. Education will be successful only when dropping out is irrational and when everyone can enjoy and use the progress of knowledge. No one knows how great the ultimate changes will have to be, but surely we know enough to begin to adjust the educational system so that everyone can better understand human nature and the nature of the world.

Credits

Chapter 1

Portions of this chapter were previously presented in *Gender and Anthropology: Critical Reviews for Research and Teaching* edited by S. Morgan (Lancaster, 1989a); in the Proceedings of the XIIIth Congress of the International Primatological Society edited by T. Nishida, W. C. McGrew, P. Marler, M. Pickford and F. de Waal (Lancaster and Kaplan, in press); and from the distinguished speaker address delivered at the 1991 Annual Luncheon of the American Association of Physical Anthropologists at meetings in Milwaukee, April 4, 1991, to be published in the *1991 Yearbook of Physical Anthropology*.

Chapter 2

The key ideas that have culminated in this chapter have been brewing for a long time. As such, their development has been influenced significantly over the years by discussion with many colleagues and students. Particularly influential in this regard have been my long-term colleagues and friends Michael J. Raleigh, Stevan Harnad, Arthur Kling, Robin Fox, and Lionel Tiger. The scheme of the proximate and ultimate causes of social systems presented in this chapter has gone through many revisions following late hour discussions with my colleague, co-conspirator, and wife, C. Netzin Gerald. The figure, and many of the ideas that flow from it, are the result of our joint intellectual effort.

Chapter 9

Xiaoming Chen, Dianne Rekow, Mark Teaford, Steve Koch, Col. Bruce Altschuler, Michael Charney, Thomas O. McCracken, Paul Newmann, and Robert Parshall provided the images used in this chapter. The archaeological specimen illustrated in figure 3 was provided by the Lowie Museum of Anthropology, University of California, Berkeley. The IBM Palo Alto Scientific Center and the L.S.B. Leakey Foundation cosponsored a "Prehistory in the Future" symposium in 1991 from which many of our ideas are derived. Robert C. Taylor, William A. Hanson, Gary Richards, and Alfred L. Rosenberger and several anonymous reviewers of an earlier version of this manuscript provided assistance and many useful suggestions.

Bibliography

Adams, L. P. and D. A. Constant. 1988. Biostereometrics in the Study of Morphology of the Lumbar Sacral Spine. *Medical Biological Engineering Computing*, 26:353–88.

Adams, L. P., C. H. Jooste and C. J. Thomas. 1989. An Indirect In Vivo Method for Quantification of Wear of Denture Teeth. *Dental Materials*, 5:31–34.

Adams, L. P. and R. J. C. Wilding. 1988. Tooth Wear Measurements Using a Reflex Microscope. *Journal of Oral Rehabilitation*, 15:605–13.

Adams, M. D., J. M. Kelley, J. D. Gocayne, M. Dubnick, M. H. Polymeropoulos, H. Xiao, C. R. Merril, A. Wu, B. Olde, R. F. Moreno, A. R. Kerlavage, W. R. McCombie and J. C. Venter. 1991. Complementary DNA Sequencing: Expressed Sequence Tags and Human Genome Project. *Science*, 252:1,651–56.

Agoramoorthy, G. and S. M. Mohnot. 1988. Infanticide and Juvenilicide in Hanuman Langurs (*Presbytis entellus*) Around Jodhpur. *Human Evolution*, 3:279–96.

Agoramoorthy, G., S. M. Mohnot, V. Sommer and A. Sriviastava. 1988. Abortions in Free Ranging Hanuman Langurs (*Presbytis entellus*): A Male Induced Strategy? *Human Evolution*, 3:297–308.

Alberius, P., M. Malmberg, S. Persson and G. Selvik. 1990. Variability of Measurements of Cranial Growth in the Rabbit. *American Journal of Anatomy*, 188:393–400.

Alexander, R. D. 1974. The Evolution of Social Behavior. *Annual Review of Ecology and Systematics*, 5:325–83.

Alexander, R. D., R. C. Hoogland, R. D. Howard, K. Noonan and P. Sherman. 1979. Sexual Dimorphisms and Breeding Systems in Pinnipeds, Ungulates, Primates and Humans. In *Evolutionary Biology and Human Social Behavior*, N. A. Chagnon and W. Irons (eds.), pp. 402–35. North Scituate, MA: Duxbury Press.

Alexander, R. D. and K. M. Noonan. 1979. Concealment of Ovulation, Parental Care, and Human Social Evolution. In *Evolutionary Biology and Human Social Behavior*, H. A. Chagnon and W. Irons (eds.), pp. 436–53. North Scituate, MA: Duxbury Press.

Altmann, J. 1987. Lifespan Aspects of Reproduction and Parental Care in Anthropoid Primates. In *Parenting Across the Life Span*, J. Lancaster, J. Altmann, A. S. Rossi and L. R. Sherrod (eds.), pp. 15–29. Hawthorne, NY: Aldine de Gruyter.

Altmann, J. 1974. Observational Study of Behavior: Sampling Methods. *Behaviour*, 49:227–67.

Altmann, S. 1967. The Structure of Primate Social Communication. In *Social Communication Among Primates*, S. Altmann (ed.), pp. 325–62. Chicago: University of Chicago Press.

Altmann, S. A. and J. Altmann. 1979. Demographic Constraints on Behavior and Social Organization. In *Primate Ecology and Human Origins: Ecological Influences on Social Organization*, I. S. Bernstein and E. O. Smith (eds.). New York: Garland Press.

Altschuler, B. R. 1976. Applications of Interferometry and Optical Metrology in Dentistry. *Society of Photo-Optical Instrumentation Engineers Proceedings*, 179:40–46.

Altschuler, B. R., J. Taboada and M. Altschuler. 1979. Imaging Applications for Automated Industrial Inspection and Assembly. *Society of Photo-Optical Instrumentation Engineers Proceedings*, 182:192–96.

Altschuler, M., J. L. Posdamer, G. Frieder, B. R. Altschuler and J. Taboada. 1981. The Numerical Stereo Camera. *Society of Photo-Optical Instrumentation Engineers Proceedings*, 183:15–24.

Anderson, P. 1983. The Reproductive Role of the Human Breast. *Current Anthropology*, 24:25–46.

Andrews, P. and J. Cronin. 1982. The Relationship of *Sivapithecus* and *Ramapithecus* and the Evolution of the Orangutan. *Nature*, 297:541–46.

Andrews, P. and J. A. Van Couvering. 1975. Palaeoenvironments in the East African Miocene. *Contributions in Primatology*, 5:62–103.

Ardrey, R. 1961. *African Genesis*. New York: Atheneum Press.

Aronson, L. R. 1959. Hormones and Reproductive Behavior: Some Phylogenetic Considerations. In *Comparative Endocrinology*, A. Gorbman (ed.), pp. 98–120. New York: John Wiley and Sons.

Barbetti, M. 1986. Traces of Fire in the Archaeological Record Before One Million Years Ago? *Journal of Human Evolution*. 15:771–81.

Barlow, G. W. 1989. Has Sociobiology Killed Ethology or Revitalized It? In *Perspectives in Ethology: Whither Ethology?*, P. P. G. Bateson and P. H. Klopfer (eds.), pp.1–45. New York: Plenum Press.

_____. 1988. Monogamy in Relation to Resources. In *The Ecology of Social Behavior*, C. N. Slobodchikoff (ed.), pp. 55–79. New York: Academic Press.

Bartholomew, G. A. and J. B. Birdsell. 1953. Ecology and the Protohominids. *American Anthropologist*, 55:481–98.

Bateson, P. 1984. The Biology of Cooperation. *New Society*, 31 (May): 343–45.

Bayne, S. 1989. CAD/CAM: Science and Technology. *Transactions of the Academy of Dental Materials: Proceedings of the Conference on CAD/CAM in Dentistry*, 2:3–7.

Beach, F. A. 1947. A Review of Physiological and Psychological Studies of Sexual Behavior in Mammals. *Physiological Reviews*, 27:240–307.

Becquerel, H. 1896. Sur les Radiations Invisibles Emises par les Corps Phosphorescents. *Comptes Rendus de l'Academie des Sciences*, 122:501–503.

Beer, C. G. 1982. Study of Vertebrate Communication: Its Cognitive Implications. In *Animal Mind—Human Mind*, D. R. Griffin (ed.), pp. 251–67. Berlin: Springer-Verlag.

245

Berger, J. 1988. Social Systems, Resources, and Phylogenetic Inertia: An Experimental Test and Its Limitations. In *The Ecology of Social Behavior*, C. N. Slobodchikoff (ed.), pp. 157–86. New York: Academic Press.

Bernstein, I. S. 1984. The Adaptive Value of Maladaptive Behavior, or You've Got to Be Stupid in Order to Be Smart. *Ethology and Sociobiology*, 5:297–303.

———. 1981. Dominance: The Baby and the Bath Water. *Behavioral and Brain Sciences*, 4:419–57.

Bernstein, I. S., T. P. Gordon and R. M. Rose. 1983. The Interaction of Hormones, Behavior and Social Context in Nonhuman Primates. In *Hormones and Aggressive Behavior*, B. B. Svare (ed.). New York: Plenum Press.

Betzig, L. B., M. Mulder and P. Turke (eds.). 1988. *Human Reproductive Behaviour: A Darwinian Perspective*. New York: Cambridge University Press.

Beynon, A. D. and M. C. Dean. 1988. Distinct Dental Development Patterns in Early Fossil Hominids. *Nature*, 335:509–14.

Biknevicius, A. R. 1986. Dental Function and Diet in the Carpolestidae (Primates, Plesiadapiformes). *American Journal of Physical Anthropology*, 71:157–71.

Bishop, N. 1975. Social Behavior of Langur Monkeys (*Presbytis entellus*) in a High Altitude Environment. Ph.D. Thesis, University of California, Berkeley.

Bishop, N., S. B. Hrdy, J. Teas and J. Moore. 1981. Measures of Human Influence in Habitats of South Asian Monkeys. *International Journal of Primatology*, 2:153–67.

Bishop, W. W. and J. A. Miller. 1972. *Calibration of Hominoid Evolution*. Edinburgh: Scottish Academic Press Ltd.

Blumenschine, R. J. 1986. Early Hominid Scavenging Opportunities: Implication of Carcass Availability in the Serengeti and Ngorongoro Eco-systems. *B.A.R. International Series*, 283.

Blumenschine, R. J. and J. A. Cavallo. 1992. Scavenging and Human Evolution. *Scientific American*, October: 90–96.

Boaz, N. T. 1990. *Evolution of Environments and Hominidae in the West African Rift Valley*. Martinsville: Virginia Museum of Natural History, Memoir 1.

———. 1988. Status of *Australopithecus afarensis*. *Yearbook of Physical Anthropology*, 31:85–113.

———. 1987. Sahabi and Neogene Hominoid Evolution. In *Neogene Paleontology and Geology of Sahabi, Libya*, N. T. Boaz, et al. (eds.), pp. 129–34. New York: Alan R. Liss.

———. 1983. Morphological Trends and Phylogenetic Relationships from Middle Miocene Hominoids to Late Pliocene Hominids. In *New Interpretations of Ape and Human Ancestry*, R. L. Ciochon and R. S. Corruccini (eds.), pp. 705–20. New York: Plenum.

Boaz, N. T., A. El-Arnauti, A. W. Gaziry, J. de Heinzelin and D. D. Boaz. 1987. *Neogene Paleontology and Geology of Sahabi Libya*. New York: Alan R. Liss.

Boaz, N. T., P. P. Pavlakis and A. Brooks. 1990. Late Pleistocene-Holocene Human Remains from the Upper Semliki, Zaire. In *Evolution of Environments and Hominidae in the African Western Rift Valley*, Noel T. Boaz (ed.), pp. 273–99. Martinsville: Virginia Museum of Natural History, Memoir 1.

Boesch, C. 1991. The Effects of Leopard Predation on Grouping Patterns in Forest Chimpanzees. *Behaviour*, 117(3–4): 220–42.

Boggess, J. E. 1980. Intermale Relations and Troop Male Membership Changes in Langurs (*Presbytis entellus*) in Nepal. *International Journal of Primatology*, 1:233–74.

_____. 1979. Troop Male Membership Change and Infant Killing in Langurs (*Presbytis entellus*). *Folia Primatologica*, 32:35–107.

Boone, J. 1988. Parental Investment, Social Subordination, and Population Processes Among the 15th and 16th Century Portuguese Nobility. In *Human Reproductive Behaviour: A Darwinian Perspective*, L. Betzig, M. B. Mulder and P. Turke (eds.), pp. 201–219. New York: Cambridge University Press.

_____. 1986. Parental Investment and Elite Family Structure in Preindustrial States: A Case Study of Late Medieval-Early Modern Portuguese Genealogies. *American Anthropologist*, 88:859–78.

Borkan, G. A., D. E. Hults, S. G. Gerzov and A. H. Robbins. 1985. Comparison of Body Composition in Middle-Aged and Elderly Males Using Computed Tomography. *American Journal of Physical Anthropology*, 66:289–95.

Bowden, D. M., G. L. Brammer, T. Rederickson, M. J. Raleigh, A. Dougherty, R. A. Short and D. D. Williams. 1989. Relations Among Whole Blood Serotonin and Sex, Age, Diet, and Social Status in *Macaca nemestrina*. *American Journal of Primatology*, 18:221–30.

Bowlby, J. 1969. *Attachment and Loss: Attachment*. London: Hogarth Press.

Bowler, P. J. 1989. *The Mendelian Revolution: The Emergence of Hereditarian Concepts in Modern Science and Society*. Baltimore: The John Hopkins University Press.

Boyd, W. C. 1950. *Genetics and the Races of Man*. Boston: D. C. Heath.

Boyde, A. 1971. Comparative Histology of Mammalian Teeth. In *Dental Morphology and Evolution*, A. A. Dahlberg (ed.), pp. 81–94. Chicago: University of Chicago Press.

_____. 1967. The Development of Enamel Structure. *Proceedings of the Royal Society of Medicine*, 60:13–18.

Brain, C. K. 1981. The Evolution of Man in Africa: Was It a Consequence of Cainozoic Cooling? *Geological Society*, 84:1–19.

_____. 1972. An Attempt to Reconstruct the Behavior of *Australopithecus*: The Evidence for Interpersonal Violence. *South African Museum Association Bulletin*, 9:127–39.

Brain, C. K., C. S. Churcher, J. D. Clark, F. E. Grine, P. Shipman, R. L. Susman, A. Turner and V. Watson. 1988. New Evidence of Early Hominids: Their Culture and Environment from the Swartkrans Cave, South Africa. *South African Journal of Science*, 84:828–35.

Brauer, G. 1989. The Evolution of Modern Humans: A Comparison of the African and Non-African Evidence. In *The Human Revolution*, P. Mellars and C. Stringer (eds.), pp. 123–54. Edinburgh: Edinburgh University Press.

Brice, W. K. 1982. Bishop Ussher, John Lightfoot and the Age of Creation. *Journal of Geological Education*, 30(1): 18–24.

Brooks, A. S. and C. C. Smith. 1987. Ishango Revisited: New Age Determinations and Cultural Interpretations. *African Archaeological Review*, 5:67–78.

Brooks, J. L. 1984. *Just Before the Origin: Alfred Russell Wallace's Theory of Evolution*. New York: Columbia University Press.

Broom, R. 1950. The Genera and Species of the South African Fossil Ape-Men. *The American Journal of Physical Anthropology*, 8:1–14.

Brothers, L. 1990. The Social Brain: A Project for Integrating Primate Behavior and Neurophysiology in a New Domain. *Concepts in Neuroscience*, 1(1): 27–51.

Brown, J. K. 1970. A Note on the Division of Labor by Sex. *American Anthropologist*, 72:1,073–78.

Brown, P. H. and G. T. Krishnamurthy. 1990. Design and Operation of a Nuclear Medicine Picture Archiving and Communication System. *Seminars in Nuclear Medicine*, 20:205–224.

Buirski, P., R. Plutchik and H. Kellerman. 1978. Sex Differences, Dominance, and Personality in the Chimpanzee. *Animal Behaviour*, 26:123–29.

Burchfield, J. D. 1975. *Lord Kelvin and the Age of the Earth*. New York: Science History Publications.

Burgess, R. and P. Draper. 1989. The Explanation of Family Violence: The Role of Biological, Behavioral, and Cultural Selection. In *Crime and Justice: A Review of Research*, Vol. 11, M. Tonry and N. Morris (eds.), pp. 59–116. Chicago: University of Chicago Press.

Burke, P. H. and C. A. Hughes-Lawson. 1989. Developmental Changes in the Facial Soft Tissues. *American Journal of Physical Anthropology*, 79:281–88.

Burton, L. M. 1990. Teenage Childbearing as an Alternative Life-Course Strategy in Multigenerational Black Families. *Human Nature*, 1:123–44.

Buss, D. M. 1991. Evolutionary Personality Psychology. *Annual Review of Psychology*, 42:459–91.

Butler, P. M. 1952. The Milk Molars of Perissodactyla, with Remarks on Molar Occlusion. *Proceedings of the Royal Society of London*, 121:777–817.

Caccone, A. and J. R. Powell. 1989. DNA Divergence Among Hominoids. *Evolution*, 43:925–42.

Cann, R. L., M. Stoneking and A. C. Wilson. 1987. Mitochondrial DNA and Human Evolution. *Nature*, 325:31–36.

Cavalli-Sforza, L. L., A. Piazza, P. Menozzi and J. Mountain. 1988. Reconstruction of Human Evolution: Bringing Together Genetic, Archaeological and Linguistic Data. *Proceedings, the National Academy of Sciences (USA)*, 85:6,002–6.

Chadwick-Jones, J. K. 1987. Social Psychology and Primatology: Proximate Explanations. *Ethology*, 74:164–69.

Charteris, J., J. C. Wall and J. Nottrodt. 1981. Functional Reconstruction of Gait from the Pliocene Hominid Footprints at Laetoli, Northern Tanzania. *Nature*, 290:496–98.

Cheney, D. L. and R. M. Seyfarth. 1990. *How Monkeys See the World: Inside the Mind of Another Species*. Chicago: University of Chicago Press.

Cheney, D. L. and R. W. Wrangham. 1987. Predation. In *Primate Societies*, B. B. Smuts, D. L. Cheney, R. M. Seyfarth, R. W. Wrangham and T. T. Struhsaker. (eds.), pp. 227–39. Chicago: The University of Chicago Press.

Ciochon, R. L. and J. G. Fleagle (eds.). 1985. *Primate Evolution and Human Origins*. Menlo Park, CA: Benjamin/Cummings.

Clark, A. B. and T. J. Ehlinger. 1987. Pattern and Adaptation in Individual Behavioral Differences. In *Perspectives in Ethology: Alternatives*, P. P. G. Bateson and P. H. Klopfer (eds.), pp. 1–47. New York: Plenum Press.

Clark, J. D. 1988. The Middle Stone Age of East Africa and the Beginnings of Regional Identity. *Journal of World Prehistory*, 2(3): 235–305.

_____. 1987. Transitions: Homo Erectus and the Acheulian: The Ethopian Sites of Gadeb and the Middle Awash. *Journal of Human Evolution*, 16(7/8): 809–26.

_____. 1970. *The Prehistory of Africa*. London: Thames and Hudson.

Clark, J. D. and J. W. K. Harris. 1985. Fire and its Roles in Early Hominid Life-Ways. *The African Archaeological Review*, 3:3–27.

Clark, J. D., M. A. J. Williams and A. B. Smith. 1973. The Geomorphology and Archaeology of Adrar Bous, Central Sahara: A Preliminary Report. *Quaternaria*, 17:245–97.

Clark, W. E. le Gros. 1971. *The Antecedents of Man*. Chicago: Quadrangle.

_____. 1964. *The Fossil Evidence for Human Evolution*, 2nd Ed. Chicago: University of Chicago Press.

_____. 1947. Observations on the Anatomy of the Fossil Australopithecinae. *Journal of Anatomy*, 81:300–34.

Clarke, A. S., W. A. Mason and G. P. Moberg. 1988. Differential Behavioral and Adrenocortical Responses to Stress Among Three Macaque Species. *American Journal of Primatology*, 14:37–52.

Clutton-Brock, T. H. and P. H. Harvey. 1978. Mammals, Resources and Reproductive Strategies. *Nature*, 273:191–95.

_____. 1977. Primate Ecology and Social Organization. *Proceedings of the Zoological Society of London, Journal of Zoology*, 183:1–39.

Conroy, G. C., M. Pickford, B. Senut, J. Van Couvering and P. Mein. 1992. *Otavipithecus namibiensis*, First Miocene Hominoid from Southern Africa, *Nature*, 356:144–47.

Conroy, G. C. and M. W. Vannier. 1987. Dental Development of the Taung Skull from Computerized Tomography. *Nature*, 329:625–27.

Conroy, G. C., M. W. Vannier. 1984. Noninvasive Three-Dimensional Computer Imaging of Matrix-Filled Fossil Skulls by High-Resolution Computed Tomography. *Science*, 226:456–58.

Conroy, G. C., M. W. Vannier and P. V. Tobias. 1990. Endocranial Features of Australopithecus Africanus Revealed by 2- and 3-D Computed Tomography. *Science*, 247:838–41.

Coombes, A. M., J. P. Moss, A. D. Linney and R. Richards. 1990. A Method for the Analysis of the 3-D Shape of the Face and Changes in the Shape Brought about by Facial Surgery. *Society of Photo-Optical Instrumentation Engineers Proceedings*, 1,380:53.

Cowlishaw, G. and R. I. M. Dunbar. 1991. Dominance Rank and Mating Success in Male Primates. *Animal Behaviour*, 41:1,045–56.

Creel, N. 1978. Stereometrics in Primate Taxonomy and Phylogeny. *Society of Photo-Optical Instrumentation Engineers Proceedings*, 166:338–45.

Creel, S., N. Creel, D. E. Wildt and S. L. Monfort. 1992. Behavioural and Endocrine Mechanisms of Reproductive Suppression in Serengeti Dwarf Mongooses. *Animal Behaviour*, 43(2): 231–46.

Crews, D. 1987. Diversity and Evolution of Behavioral Controlling Mechanisms. In *Psychobiology of Reproductive Behavior: An Evolutionary Perspective*, D. Crews (ed.), pp. 88–119. Englewood Cliffs, NJ: Prentice-Hall.

Crews, D. and M. C. Moore. 1986. Evolution of Mechanisms Controlling Mating Behavior. *Science*, 231:121–230.

Curtin, R. A. 1981. Strategy and Tactics in Male Grey Langur Competition. *Journal of Human Evolution*, 10:245–53.

_____. 1977. Langur Social Behavior and Infant Mortality. *Berkeley Papers in Anthropology*, 50:27–36.

_____. 1975. The Socioecology of the Common Langur (*Presbytis entellus*) in the Nepal Himalaya. Ph.D. Thesis, University of California, Berkeley.

Curtin, R. A. and P. Dolhinow. 1978. Primate Social Behavior in a Changing World. *American Scientist*, 66:468–75.

Curtis, G. 1981. *A Guide to Dating Methods for the Determination of the Last Time of Movement of Faults*. For the United States Nuclear Regulatory Commission, Office of Nuclear Regulatory Research. Washington, DC: Government Printing Office.

_____. 1975. Age of KBS Tuff in Koobi Fora Formation, East Rudolf, Kenya. *Nature*, 258:395.

Cutting, C. B., F. L. Bookstein, B. J. Grayson, L. Fellingham and J. G. McCarthy. 1986. Three-Dimensional Computer-Assisted Design of Craniofacial Surgical Procedures: Optimization and Interaction with Cephalometric and CT-Based Models. *Plastic and Reconstructive Surgery*, 77:877–85.

Daegling, D. J. 1989. Biomechanics of Cross-Sectional Size and Shape in the Hominoid Mandibular Corpus. *American Journal of Physical Anthropology*, 80:91–106.

Dahlberg, A. A. and W. G. Kinzey. 1962. Etude Microscopique de l'Abrasion de l'Attrition sur la Surface des Dents. *Bulletin of Group International Research Scientists in Stomatology and Odontology*, 21:36–60.

Dahlberg, F. (ed.) 1981. *Woman The Gatherer*. New Haven: Yale University Press.

Daly, M. and M. Wilson. 1988. *Homicide*. Hawthorne, NY: Aldine de Gruyter.

_____. 1982. Male Sexual Jealousy. *Ethology and Sociobiology*, 3:11–27.

Damasio, A. and H. Damasio. 1992. Brain and Language. *Scientific American*, 267(3): 89–95.

Dart, R. A. 1957a. The Makapansgat Australopithecine Osteodontokeratic Culture. *Proceedings of the 3rd Pan-African Congress on Prehistory*: 161–71.

_____. 1957b. *The Osteodontokeratic Culture of* Australopithecus africanus. Pretoria: Transvaal Museum Memoirs 10.

_____. 1948. The Makapansgat Proto-Human *Australopithecus prometheus*. *American Journal of Physical Anthropology*, 6:259–83.

_____. 1925. *Australopithecus africanus*: The Man-Ape of South Africa. *Nature*, 115:195–96.

Darwin C. 1871. *The Descent of Man and Selection in Relation to Sex*, 2nd Ed. London: C. A. Watts and Company.

_____. 1859. *On the Origin of Species*. London: James Murray.

Davies, N. B. and A. Lundberg. 1984. Food Distribution and a Variable Mating System in the Dunnock, *Prunella modularis*. *Journal of Animal Ecology*, 53:895–912.

Dawkins, M. S. 1989. The Future of Ethology: How Many Legs Are We Standing On? In *Perspectives in Ethology: Whither Ethology?*, P. P. G. Bateson and P. H. Klopfer (eds.), pp. 47–54. New York: Plenum Press.

Dawkins, R. 1976. *The Selfish Gene*. Oxford: Oxford University Press.

Dayhoff, M. 1972. *Atlas of Protein Sequence and Structure*, Vol 5, p. 15. Washington, DC: Georgetown University.

Dean, D. 1981. Age of the Earth Controversy. *Annals of Science*, 38:435–56.

de Heinzelin, J. 1955. *Le Fossè Tectonique sous le parallel d'Ishango*. *Exploration Parc Nat. Albert, Mission J. de Heinzelin de Braucourt* (1950), fasc. 1. Brussels: Inst. Parcs Nat. Congo Belge.

de Heinzelin, J. and A. El-Arnauti. 1987. The Sahabi Formation and Related Deposits. In *Neogene Paleontology and Geology of Sahabi, Libya*, N. T. Boaz, et al. (eds.), pp. 1–21. New York: Alan R. Liss.

Deino, A. and R. Potts. 1990. Single-Crystal 40Ar/39Ar Dating of the Olorgesailie Formation, Southern Kenya Tuft. *Journal of Geophysical Research*, 95:8,453–70.

Delson, E. 1988. Chronology of South African Australopith Site Units. In *Evolutionary History of the "Robust" Australopithecines*, Fred Grine (ed.), pp. 317–24. New York: Aldine di Gruyter.

Delson, E., I. Tattersall and J. Van Couvering. 1991. *Paleoanthropology Annuals*, Vol. 1 (1990). New York: Garland.

Dennell, R. W., H. M. Rendell and E. Hailwood. 1988. Late Pliocene Artifacts from Northern Pakistan. *Current Anthropology*, 29:495–98.

Desio, A. 1935. Appunti Geologici sui Dintorni di Sahabi (Sirtica). *Rendiconti Reale Istituto Lettere*, ser. 2, vol. 68, fasc. 1-V:137–44.

DeVore, I. 1965. Male Dominance and Mating Behavior in Baboons. In *Sex and Behavior*, F. Beach (ed.), pp. 266–89. New York: John Wiley and Sons.

DeVore, I. and S. L. Washburn. 1963. Baboon Ecology and Human Evolution. In *African Ecology and Human Evolution*, F. C. Howell (ed.), pp. 335–67. Chicago: Aldine Press.

de Waal, F. B. M. 1989. Dominance 'Style' and Primate Social Organization. In *Comparative Socioecology: The Behavioural Ecology of Humans and Other Mammals*, V. Standen and R. A. Foley (eds.), pp. 243–63. London: Blackwell.

de Waal, F. B. M. and L. M. Luttrell. 1989. Toward a Comparative Socioecology of the Genus *Macaca*: Different Dominance Styles in Rhesus and Stumptail Monkeys. *American Journal of Primatology*, 19:83–109.

Dickemann, M. 1981. Paternity Confidence and Dowry Competition: A Biocultural Analysis of Purdah. in *Natural Selection and Social Behavior*, R. Alexander and D. Tinkle (eds.), pp. 417–38. New York: Cheiron Press.

––––––. 1979a. Female Infanticide, Reproductive Strategies and Social Stratification: A Preliminary Model. In *Evolutionary Biology and Human Social Behavior*, N. Chagnon and W. G. Irons (eds.), pp. 321–67. North Scituate, MA: Duxbury Press.

––––––. 1979b. The Ecology of Mating Systems in Hypergynous Dowry Societies. *Social Science Information*, 18:163–95.

Dittrich, W. 1990. Representation of Faces in Longtailed Macaques (*Macaca fascicularis*). *Ethology*, 85:265–78.

Dobbing, J. 1974. The Later Development of the Brain and its Vulnerability. In *Scientific Foundations of Pediatrics*, J. A. Davis and J. Dobbing (eds.), pp. 565–77. Philadelphia: W. B. Saunders.

Dobzhansky, T., F. J. Ayala, G. L. Stebbins and J. W. Valentine. 1977. *Evolution*. San Francisco: W. H. Freeman.

Dolhinow, P. (ed.). 1978. *Primate Patterns*. New York: Holt, Rinehart and Winston.

––––––. 1972. *Primate Patterns*. New York: Holt, Rinehart and Winston.

Dolhinow, P. and M. Taff. 1990. Changing Social Relations in a Captive All-Male Group of Langur Monkeys (*Presbytis entellus*). *American Journal of Physical Anthropology*, 81(2): 305 (abstract only).

Doolittle, R. F. 1987. *Of URFS and ORFS*, p. 33. Mill Valley, CA: University Science Books.

Draper, P. In press. Room to Maneuver: !Kung Women Cope with Men. In *Sanctions and Sanctuary: Cultural Perspectives on the Beating of Wives*, D. Counts, J. K. Brown and J. Campbell (eds.). Boulder: Westview Press.

Dunbar, R. I. M. 1989. Reproductive Strategies of Female Gelada Baboons. In *The Sociobiology of Reproductive Strategies*, A. E. Rasa, C. Vogel and E. Voland (eds.), pp. 74–282. New York: Chapman & Hall.

––––––. 1988. *Socio-Ecological Systems in Primate Social Systems*, ch. 12, pp. 262–91. Ithaca: Cornell University Press.

Dunbar, R. I. M. 1986. The Social Ecology of Gelada Baboons. In *Ecological Aspects of Social Evolution: Birds and Mammals*, D. I. Rubenstein and R. W. Wrangham (eds.), pp. 332–51. Princeton: Princeton University Press.

Duret, F., J. L. Bloin and B. Duret. 1988. CAD-CAM in Dentistry. *Journal of the American Dental Association*, 117:715–20.

Edey, M. A. 1972. *The Missing Link*. New York: Time-Life.

Eglinton, G. and G. B. Curry (eds.). 1991. Molecules Through Time: Fossil Molecules and Biochemical Systematics. *Philosophical Transactions of the Royal Society of London, Series B*, 333:315–433.

Emlen, S. and L. Oring. 1977. Ecology, Sexual Selection and the Evolution of Mating Systems. *Science*, 197:215–23.

Erlich, H. (ed.). 1989. *PCR Technology*. New York: Stockton Press.

Essock-Vitale, S. and R. M. Seyfarth. 1987. Intelligence and Social Cognition. In *Primate Societies*, B. B. Smuts, D. Cheney, R. M. Seyfarth, R. W. Wrangham and T. T. Struhsaker (eds.), pp. 452–61. Chicago: The University of Chicago Press.

Ewert, J. P., R. R. Capranica and D. J. Ingle. 1983. *Advances in Vertebrate Neuroethology*. New York: Plenum Press.

Falk, D., J. M. Cheverud, M. W. Vannier and G. C. Conroy. 1986. Advanced Computer Graphics Technology Reveals Cortical Assymetry in Endocasts of Rhesus Monkeys. *Folia Primatologica*, 46:98–103.

Faure, G. 1986. *Principles of Isotope Geology*, 2nd Ed. New York: John Wiley and Sons.

Fedigan, L. M. 1986. The Changing Role of Women in Models of Human Evolution. *Annual Review of Anthropology*, 15:22–66.

_____. 1982. *Primate Paradigms: Sex Roles and Social Bonds*. Montreal: Eden Press.

Fink, W. 1990. Data Acquisition in Systematic Biology. In *Proceedings of the Michigan Morphometrics Workshop*, F. J. Rohlf and F. L. Bookstein (eds.), pp. 9–20. Ann Arbor: University of Michigan Museum of Zoology.

Fleischer, R. L., R. M. Price, R. M. Walker and L. S. B. Leakey. 1965. Fission-Track Dating of Bed I, Olduvai Gorge. *Science*, 148:72.

Fleming, S. J. 1979. *Thermoluminescence Techniques in Archaeology*. Oxford: Clarendon Press.

Foley, R. A. and P. C. Lee. 1989. Finite Social Space, Evolutionary Pathways, and Reconstructing Hominid Behavior. *Science*, 243:901–6.

Gardner, A. and B. Gardner. 1969. Teaching Sign Language to a Chimpanzee. *Science*, 165:664–72.

Gardner, A., B. Gardner and T. Van Cantfort. 1989. *Teaching Sign Language to Chimpanzees*. Albany: State University of New York Press.

Gaulin, S. J. C. 1979. A. Jarman/Bell Model of Primate Feeding Niches. *Human Ecology*, 7(1): 1–20.

Gaulin. S. J. C. and R. W. Fitzgerald. 1989. Sexual Selection for Spatial-Learning Ability. *Animal Behaviour*, 37:322–31.

Gaulin, S. J. C. and L. D. Sailer. 1985. Are Females the Ecological Sex? *American Anthropologist*, 87:111–19.

Gaulin, S. J. C. and A. Schlegel. 1980. Paternal Confidence and Paternal Investment: A Cross-Cultural Test of a Sociobiological Hypothesis. *Ethology and Sociobiology*, 1:301–9.

Gautier, A. 1965. *Geological Investigation in the Sinda-Mohari (Ituri, NE Congo): A Monograph on the Geological History of a Region in the Lake Albert Rift*. Gent, Belgium: Rijksuniv, Gent.

Geikie, A. 1905. *Founders in Geology*. London: Macmillan.

Geronimus, A. T. 1987. On Teenage Childbearing and Neonatal Mortality in the United States. *Population Development Review*, 13:245–79.

_____. 1986. The Effects of Race, Residence, and Prenatal Care on the Relationship of Maternal Age to Neonatal Mortality. *American Journal of Public Health*, 76:1,416–21.

Geyh, M. and H. Schleicher. 1990. *Absolute Age Determination: Physical and Chemical Dating Methods and Their Application*. Berlin: Springer-Verlag.

Goodall, J. 1986. *The Chimpanzees of Gombe: Patterns of Behavior*. Cambridge, MA: Bellknap Press.

Gordon, K. D. 1982. A Study of Microwear on Chimpanzee Molars: Implications for Dental Microwear Analysis. *American Journal of Physical Anthropology*, 59:195–215.

Gore, A. 1989. National High-Performance Computer Technology Act of 1989: Hearings Before the Committee on Commerce, Science, and Transportation. S. 1,067:24–38.

Gore, M. A. 1991. A Comparative Study of the Relationships and Behaviours in a Female and a Non-Female Bonded Social System. In *Primatology Today*, A. Ehara, T. Kimura, O. Takenaka and M. Iwamoto (eds.), pp. 189–92. New York: Elsevier Science Publishers.

Greenfield, L. O. 1980. A Late Divergence Hypothesis. *American Journal of Physical Anthropology*, 50:527–48.

Greenwood, P. J. 1980. Mating Systems, Philopatry, and Dispersal in Birds and Mammals. *Animal Behaviour*, 28:1,140–62.

Gregory, W. K. 1949. The Bearing of the Australopithecinae Upon the Problem of Man's Place in Nature. *American Journal of Physical Anthropology*, 7:485–512.

Grine, F. E. (ed.) 1988. *Evolutionary History of the "Robust" Australopithecines*. New York: Aldine de Gruyter.

_____. 1986. Dental Evidence for Dietary Differences in Australopithecus and Paranthropus: A Quantitative Analysis of Permanent Molar Microwear. *Journal of Human Evolution*, 15:783–822.

Grine, F. E., E. E. Colflesh, D. J. Daegling, D. W. Krause, M. M. Dewey, R. H. Cameron and C. K. Brain. 1989. Electron Probe X-ray Microanalysis of Internal Structures in a Fossil Hominid Mandible and its Implications for Biomechanical Modelling. *Suid-Afrikaanse Tydskrif vir Wetenskap*, 85:509–14.

Grine, F. E. and R. F. Kay. 1988. Early Hominid Diets from Quantitative Image Analysis of Dental Microwear. *Nature*, 333:765–68.

Grun, R., H. Schwarcz and S. Zymela. 1987. Electron Spin Resonance Dating of Tooth Enamel. *Canadian Journal of Earth Science*, 24:1,022–37.

Haeckel, E. 1903. *The Evolution of Man: A Popular Exposition of the Principle Points of Human Ontogeny and Phylogeny*. New York: Appleton-Century Crofts.

Haldane, J. B. S. 1956. The Argument from Animals to Man: An Examination of its Validity for Anthropology. *Journal of the Royal Anthropological Institute of Great Britain*, 118:851–59.

Hall, R. L. 1985. *Sexual Dimorphism in* Homo Sapiens. New York: Praeger.

Hallam, A. 1973. *A Revolution in the Earth Sciences: From Continental Drift to Plate Tectonics.* Oxford: Clarendon Press.

Hamburg, D., S. Washburn and N. Bishop. 1974. Social Adaptation in Nonhuman Primates. In *Coping and Adaptation*, G. Coelho, D. Hamburg and J. Adams (eds.), pp. 3–12. New York: Basic Books.

Hanson, K. M. 1985. Image Processing: Mathematics, Engineering, or Art? *Application of Optical Instrumentation in Medicine*, 13:70–81.

Harcourt, A. H. and F. B. M. de Waal (eds.). 1992. *Coalitions and Alliances in Humans and Other Animals.* Cambridge: Oxford University Press.

Harding, R. S. O. and G. Teleki (eds.). 1981. *Omnivorous Primates: Gathering and Hunting in Human Evolution.* New York: Columbia University Press.

Harraway, D. J. 1991. *Simians, Cyborgs, and Women.* New York: Routledge.

Harrell, B. B. 1981. Lactation and Menstruation in Cultural Perspective. *American Anthropologist*, 83:796–823.

Harris, J. W. K., P. G. Williamson, J. Verniers, M. Tappen, K. Stewart, D. Helgren, J. de Heinzelin, N. Boaz and R. Bellomo. 1987. Late Pliocene Hominid Occupation in Central Africa: The Setting, Context, and Character of the Senga 5A Site, Zaire. *Journal of Human Evolution*, 16:701–28.

Hartl, D. L. and A. G. Clark. 1989. Principles of Population Genetics, 2nd Ed. Sunderland, MA: Sinauer Associates, p. 424.

Hartman, S. E. 1989. Stereophotogrammetric Analysis of Occlusal Morphology of Extant Hominoid Molars: Phenetics and Function. *American Journal of Physical Anthropology*, 80:145–66.

Harvey, P. H. and J. R. Krebs. 1990. Comparing Brains. *Science*, 249:140–46.

Harvey, P. H., R. D. Martin and T. H. Clutton-Brock. 1987. Life Histories in Comparative Perspective. In *Primate Societies*, B. B. Smuts, D. L. Cheney, R. M. Seyfarth, R. W. Wrangham and T. T. Stuhsaker (eds.), pp. 181–96. Chicago: The University of Chicago Press.

Haubitz, B., M. Prokop, W. Dohring, J. H. Ostrom and P. Wellnhofer. 1988. Computer Tomography of Archaeopteryx. *Paleobiology*, 14:206–13.

Hausfater, G. and S. B. Hrdy. 1984. *Infanticide: Comparative and Evolutionary Perspectives.* New York: Aldine.

Hay, R. 1976. *Geology of Olduvai Gorge.* Berkeley: University of California Press.

Hayes, K. J. and C. H. Nissen. 1971. Higher Mental Functions of a Home-Raised Chimpanzee. In *Behavior of Nonhuman Primates, IV*, A. M. Schrier and F. Stollnitz (eds.), pp. 60–115. New York: Academic Press.

Herman, L. and P. Morrel-Samuels. 1990. Knowledge Acquisition and Asymmetry Between Language Comprehension and Production: Dolphins and Apes as General Models for Animals. In *Interpretation and Explanation in the Study of Animal Behavior.* Vol. 1: *Interpretation, Intentionality and Communication*, M. Bekoff and D. Jamieson (eds.), pp. 283–312. Boulder: Westview Press.

Herron, R. E. 1972. Biostereometric Measurement of Body Form. *Yearbook of Physical Anthropology*, 16:80–121.

Higley, J. D. and S. J. Suomi. 1989. Temperamental Reactivity in Non-Human Primates. In *Temperament in Childhood*, G. A. Kohnstamm, J. E. Bates and M. K. Rothbart (eds.), pp. 153–67. New York: John Wiley and Sons.

Hildebolt, C. F., G. Bate and G. C. Conroy. 1986. The Microstructure of Dentine in Taxonomic and Phylogenetic Studies. *American Journal of Physical Anthropology*, 70:39–46.

Hildebolt, C. F. and M. W. Vannier. 1988. 3-D Measurement Accuracy of Skull Surface Landmarks. *American Journal of Physical Anthropology*, 76:497–503.

Hildebolt, C. F., M. W. Vannier and R. H. Knapp. 1990. Validation Study of Skull Three-Dimensional Computerized Tomography Measurements. *American Journal of Physical Anthropology*, 82:283–94.

Hill, A. and S. Ward. 1988. Origin of the Hominidae: The Record of African Large Hominoid Evolution Between 14 my and 4 my. *Yearbook of Physical Anthropology*, 31:49–83.

Hill, K. and H. Kaplan. 1988. Tradeoffs in Male and Female Reproductive Strategies among the Ache: Parts 1 and 2. In *Human Reproductive Behaviour: A Darwinian Perspective*, L. Betzig, M. B. Mulder and P. Turke (eds.), pp. 277–90, 291–306. New York: Cambridge University Press.

Hillyard, S. A. and F. E. Bloom. 1982. Brain Functions and Mental Processes. In *Animal Mind—Human Mind*, D. R. Griffin (ed.), pp. 177–200. Berlin: Springer-Verlag.

Hinde, R. A. 1990. The Interdependence of the Behavioural Sciences. *Philosophical Transactions of the Royal Society of London*, 329:217–27.

_____. 1987. *Individuals, Relationships, and Culture: Links Between Ethology and the Social Sciences*. Cambridge: Cambridge University Press.

_____. 1983. *Primate Social Relationships: An Integrated Approach*. Sunderland, MA: Sinauer Associates.

_____. 1978. Dominance and Role: Two Concepts with Dual Meaning. *Journal of Social Biological Structures*, 1:27–38.

Hofland, P. L., F. P. Ottes, A. M. Vossepoel, H. M. Kroon and L. J. Schulze-Kool. 1990. Medical Imaging Workstation: A Software Environment. *Medical Information*, 15:15–19.

Holmquist, R., M. M. Miyamoto and M. Goodman. 1988. Higher-Primate Phylogeny—Why Can't We Decide? *Molecular Biology and Evolution*, 5:201–16.

Houtermans, F. G. 1966. History of the K/Ar-Method of Geochronology. In *Potassium Argon Dating*, O. A. Schaeffer and J. Zahringer (eds.). New York: Springer-Verlag.

Hrdy, S. B. In press. The Absence of Estrus in *Homo Sapiens*. In *The Origins of Humanness*, A. Brooks (ed.). Washington, DC: Smithsonian Institution Press.

_____. 1990. Sex Bias in Nature and in History: A Late 1980's Reexamination of the 'Biological Origins' Argument. *Yearbook of Physical Anthropology*, 33:25–37.

Hrdy, S. B. 1981. *The Woman that Never Evolved.* Cambridge: Harvard University Press.

_____. 1977. *The Langurs of Abu.* Cambridge: Harvard University Press.

_____. 1974. Male-Male Competition and Infanticide Among Langurs, *Presbytis entellus,* of Abu, Rajasthan. *Folia Primatologica,* 22:19–58.

Hrdy, S. B. and P. L. Whitten. 1987. Patterning Sexual Activity. In B. B. Smuts, D. L. Cheney, R. M. Seyfarth, R. W. Wrangham and T. T. Struhsaker (eds.), pp. 370–84. Chicago: University of Chicago Press.

Hrdy, S. B. and G. Williams. 1983. Behavioral Biology and the Double Standard. In *Social Behavior of Female Vertebrates,* S. Wasser (ed.), pp. 3–17. New York: Academic Press.

Hsü, K. J., et al. 1977. History of the Mediterranean Salinity Crisis. *Nature,* 267:399–403.

Huijsmans, D. P., W. H. Lamers, J. A. Los and J. Strackee. 1986. Toward Computerized Morphometric Facilities: A Review of 58 Software Packages for Computer-Aided Three-Dimensional Reconstruction, Quantification, and Picture Generation from Parallel Serial Sections. *Anatomical Record,* 216:449–70.

Hurtado, M., K. R. Hill and H. Kaplan. 1992. Women and Work: Trade-off Decisions Between Production and Child Care. *Human Nature,* 3(2).

Huss-Ashmore, R. 1980. Fat and Fertility: Demographic Implication of Differential Fat Storage. *Yearbook of Physical Anthropology,* 23:65–91.

Huxley, T. H. 1863. *Evidence as to Man's Place in Nature.* London: Williams and Norgate.

Imasato, Y. 1985. Optical Disk Archiving and Storage System. *British Journal of Radiology,* 58:802.

Ims, R. A. 1988. Spatial Clumping of Sexually Receptive Females Induces Space Sharing Among Male Voles. *Nature,* 335:541–43.

Irons, W. 1988. Parental Behaviour in Humans. In *Human Reproductive Behaviour,* L. Betzig, M. Borgerhoff Mulder and P. Turke (eds.), pp. 307–16. New York: Cambridge University Press.

_____. 1983. Human Female Reproductive Strategies. In *Social Behavior of Female Vertebrates,* S. Wasser (ed.), pp. 169–213. New York: Academic Press.

_____. 1979. Cultural and Biological Success. In *Evolutionary Biology and Human Social Behavior,* N. A. Chagnon and W. Irons (eds.), pp. 257–72. North Scituate, MA: Duxbury Press.

Isaac, G. L. 1978. The Foodsharing Behavior of Protohuman Hominids. *Scientific American,* 238(4): 171–88.

Isaac, G. L., J. W. K. Harris and D. Crader. 1976. Archaeological Evidence from the Koobi Fora Formation. In *Earliest Man and Environments in the Lake Rudolf Basin,* F. C. Howell, G. L. Isaac and R. E. F. Leakey (eds.), pp. 533–51. Chicago: University of Chicago Press.

Isbell, L. A. 1991. Contest and Scramble Competition: Patterns of Female Aggression and Ranging Behavior Among Primates. *Behavioral Ecology,* 2(2): 143–55.

Isbell, L. A., D. L. Cheney and R. M. Seyfarth. 1991. Group Fusions and Minimum Group Sizes in Vervet Monkeys (*Cercopithecus aethiops*). *American Journal of Primatology,* 25:57–65.

Jacobs, L. F., S. J. C. Gaulin, D. F. Sherry and G. E. Hoffman. 1990. Evolution of Spatial Cognition: Sex-Specific Patterns of Spatial Behavior Predicts Hippocampal Size. *Proceedings of the National Academy of Sciences USA*, 87:6,349–52.

Jamieson, I. G. 1986. The Functional Approach to Behavior: Is it Useful? *American Naturalist*, 172(2): 195–208.

Janson, C. H. and C. P. van Schaik. 1988. Recognizing the Many Faces of Primate Food Competition: Methods. *Behaviour*, 105(1–2): 165–86.

Jay, P. 1963. The Social Behavior of the Langur Monkey. Ph. D. Dissertation, University of Chicago.

Johanson, D. C. 1974. The Quest for Early Man in Ethiopia. *Explorer*, 16:4–11.

Johanson, D. C. and T. D. White. 1979. A Systematic Assessment of Early African Hominids. *Science*, 203:321–30.

Johnston, T. D. 1982. Selective Costs and Benefits in the Evolution of Learning. *Advances in the Study of Behavior*, 12:65–106.

Jolly, A. 1985. *Evolution of Primate Behavior*. New York: Macmillan.

Jost, R. G. and N. J. Mankovich. 1988. Digital Archiving Requirements and Technology. *Investigative Radiology*, 23:803–9.

Jukes, T. H. 1966. *Molecules and Evolution*. New York: Columbia University Press.

_____. 1965. The Genetic Code, II. *American Scientist*, 53:477–87.

Jungers, W. L. 1982. Lucy's Limbs: Skeletal Allometry and Locomotion in *Australopithecus afarensis*. *Nature*, 297:676–78.

Juni, J. E. 1990. NucNet: A Single-Vendor Clinical Picture Archiving and Communication System—Description and Reflections. *Seminars in Nuclear Medicine*, 20:193–204.

Kavanau, J. L. 1990. Conservative Behavioural Evolution, the Neural Substrate. *Animal Behaviour*, 39:758–67.

Kay, R. F. 1987. Analysis of Primate Dental Microwear Using Image-Processing Techniques. *Scanning Microscopy*, 1:657–62.

Kay, R. F. and F. E. Grine. 1989. Tooth Morphology, Wear and Diet in Australopithecus and Paranthropus from Southern Africa. In *Evolutionary History of the "Robust" Australopithecines*, F. E. Grine (ed.), pp. 427–47. New York: Aldine de Gruyter.

Keeley, L. H. and M. Toth. 1981. Microwear Polishes on Early Stone Tools from Koobi Fora, Kenya. *Nature*, 293:463–65.

Keith, Sir Arthur. 1931. *N32 Discoveries Relating to the Antiquity of Man*. London: Williams and Norgate.

Kelvin, W. T. 1871. On Geological Time. *Transactions of the Geological Society of Glasgow*, 3:1–28 and *Popular Lectures*, 2:10–64.

_____. 1852. On the Universal Tendency in Nature to the Dissipation of Mechanical Energy. *Philadelphia Magazine*, Ser. 4, 4:304–6 and *Mathematical Papers*, 1:511–14.

Kenny, G. C. 1979. An Optical Disk Replaces 25 Magnetic Tapes. *Institute of Electrical and Electronics Engineers, Spectrum*, 16:33–38.

Kimura, M. 1988. Thirty Years of Population Genetics with Dr. Crow. *Japanese Journal of Genetics*, 63:1–10.

_____. 1968. Evolutionary Rate at the Molecular Level. *Nature*, 217:624–26.

Kimura, M. and J. F. Crow. 1964. The Number of Alleles that Can Be Maintained in a Finite Population. *Genetics*, 49:725–38.

King, J. L. and T. Jukes. 1969. Non-Darwinian Evolution. *Science*, 164:788–98.

Klein, R. G. 1989. *The Human Career: Human Biological and Cultural Origins.* Chicago: University of Chicago Press.

Kling, A., J. B. Lancaster and J. Benitone. 1970. Amygdalectomy in the Free-Ranging Vervet (*Cercopithecus aethiops*). *Journal of Psychiatric Research*, 7:191–99.

Klosterman, L. L., J. T. Murai and P. K. Siiteri. 1986. Cortisol Levels, Binding, and Properties of Corticosteroid-Binding Globulin in the Serum of Primates. *Endocrinology*, 118(1): 424–34.

Kocher, T. D. and A. C. Wilson. 1991. Sequence Evolution of Mitochondrial DNA in Humans and Chimpanzees: Control Region and a Protein Coding Region. In *Evolution of Life: Fossils, Molecules, and Culture*, Osawa and Honjo (eds.), pp. 391–413. Tokyo: Springer-Verlag.

Koenig, W. D. 1988. Internal Migration in the Contemporary United States: Comparison of Measures and Partitioning of Stages. *Human Biology*, 60(6): 927–44.

Kohl-Larson, L. 1943. *Auf den Spuren des Vormenschen. Forschungen, Fahrten, und Erlebnisse in Deutsch-Ostafrika.* (Deutsch Africa-Expedition 1934–1939). Stuttgart: Strecker und Schroder.

Kortlandt, A. 1972. *New Perspectives on Ape and Human Evolution.* Amsterdam: Stichting voor Psychobiologie.

Krebs, J. R. 1990. Food-Storing Birds: Adaptive Specialization in Brain and Behaviour? *Philosophical Transactions of the Royal Society of London*, 329:153–60.

Kruuk, H. 1972. *The Spotted Hyena: A Study of Predation and Social Behavior.* Chicago: University of Chicago Press.

Kummer, H. 1978. On the Value of Social Relationships in Nonhuman Primates: A Heuristic Scheme. *Social Science Information*, 17(4–5): 687–705.

Kummer, H., V. Dasser and P. Hoynigen-Huene. 1990. Exploring Primate Social Cognition: Some Critical Remarks. *Behaviour*, 112(1–2): 84–98.

Lamarck. 1971 (original 1809). The Influence of Circumstances. In *Lamarck to Darwin: Contributions to Evolutionary Biology, 1809–1859*, H. Lewis McKinney (ed.). Lawrence, KS: Coronado Press.

Lancaster, J. B. 1989a. Women in Biosocial Perspective. In *Gender and Anthropology: Critical Reviews for Research and Teaching*, S. Morgen (ed.), pp. 95–115. Washington, DC: American Anthropological Association.

_____. 1989b. Evolutionary and Cross-Cultural Perspectives on Single-Parenthood. In *Interfaces in Psychology: Sociobiology and the Social Sciences*, R. W. Bell and N. J. Bell (eds.), pp. 63–72. Lubbock: Texas Tech University Press.

_____. 1986. Human Adolescence and Reproduction: An Evolutionary Perspective. In *School-Age Pregnancy and Parenthood: Biosocial Perspectives*, J. B. Lancaster and B. Hamburg (eds.), pp. 17–37. Hawthorne, NY: Aldine de Gruyter.

Lancaster, J. B. 1985. Evolutionary Perspectives on Sex Differences in the Higher Primates. In *Gender and the Life Course*, A. S. Rossi (ed.), pp. 3–27. Hawthorne, NY: Aldine.

Lancaster, J. B. and H. Kaplan. In press. Human Mating and Family Formation Strategies: The Effects of Variability Among Males in Quality and of the Allocation of Mating Effort and Parental Investment. In *XIIIth Congress of the International Primatological Society, Volume I, Human Origins*, T. Nishida, W. C. McGrew, P. Marler, M. Pickford and F. de Waal (eds.). Tokyo: University of Tokyo Press.

_____. 1991. The Distribution of Male Parental Investment Between Direct Descendants, Kin and Nonkin by Albuquerque Men. Paper presented at the annual meetings of the American Association of Physical Anthropologists. Milwaukee, April 1991.

Lancaster, J. B. and C. S. Lancaster. 1987. The Watershed: Change in Parental-Investment and Family-Formation Strategies in the Course of Human Evolution. In *Parenting Across the Life Span: Biosocial Dimensions*, J. B. Lancaster, J. Altmann, A. S. Rossi and L. R. Sherrod (eds.), pp. 187–205. Hawthorne, NY: Aldine de Gruyter.

_____. 1983. Parental Investment: The Hominid Adaptation. In *How Humans Adapt: A Biocultural Odyssey*, D. Ortner (ed.), pp. 33–66. Washington, DC: Smithsonian Institution Press.

Laws, J. W. and J. Vonder Haar Laws. 1984. Social Interactions Among Adult Male Langurs (*Presbytis entellus*) at Rajaji Wildlife Sanctuary. *International Journal of Primatology*, 5:31–50.

Leakey, L. S. B. 1959. A New Fossil Skull from Olduvai. *Nature*, 184:491–93.

_____. 1951. *Olduvai Gorge*. Cambridge: Cambridge University Press.

Leakey, M. D. 1984. *Disclosing the Past*. New York: Doubleday and Company.

_____. 1979. Footprints in the Ashes of Time. *National Geographic*, 155:446–57.

_____. 1978. Olduvai Gorge 1911–75: A History of the Investigations. In *Geological Background to Fossil Man: Recent Research in the Gregory Rift Valley, East Africa*, W. W. Bishop (ed.). Edinburgh: Scottish Academic Press.

_____. 1971. *Olduvai Gorge: Excavations in Beds I and II, 1960–1963*, Volume 3. Cambridge: Cambridge University Press.

Leakey, M. D. and J. M. Harris. 1987. *Laetoli: A Pliocene Site in Northern Tanzania*. Oxford: Clarendon press.

Leakey, M. D. and R. L. Hay. 1979. Pliocene Footprints in the Laetolil Beds at Laetoli, Northern Tanzania. *Nature*, 278:317–23.

Leakey, M. D., R. L. Hay, G. H. Curtis, R. E. Drake, M. K. Jackes and T. D. White. 1978. Fossil Hominids from the Laetolil Beds, Tanzania. In *Geological Background to Fossil Man: Recent Research in the Gregory Rift Valley, East Africa*, W. W. Bishop (ed.), pp. 157–70. Edinburgh: Scottish Academic Press.

Lee, R. B. 1979. *The !Kung San: Men, Women, and Work in a Foraging Society*. Cambridge: Cambridge University Press.

LeFloch-Prigent, P. 1989a. Scannographie du Crane de Petralona: Coupes Systematiques dans les Drois Plans: Premiere Partie: Resoltats Morphologiques. *Comptes Rendus de l'Academie des Sciences, Paris,* 309:1,855–62.

_____. 1989b. Scannographie du Crane de Petralona, 2 Partie: Morphometrie et Discussion. *Comptes Rendus de l'Academie des Sciences, Paris,* 309:1,997–2,003.

Leinfelder, K. F., B. P. Isenberg and M. E. Essig. 1989. A New Method for Generating Ceramic Restorations: A CAD-CAM System. *Journal of the American Dental Association,* 118:703–7.

Levy, J. 1982. Mental Processes in the Nonverbal Hemisphere. In *Animal Mind—Human Mind,* D. R. Griffin (ed.), pp. 57–73. Berlin: Springer-Verlag.

Lieberman, P. and E. S. Crelin. 1971. On the Speech of Neanderthal Man. *Linguistic Inquiry,* 2:203–22.

Linnaeus, C. 1758. *Systema naturae per regna tria naturae, secundum classes, ordines, genera, species cum characteribus, differentiis, synonymis, locis.* Edito decima, reformata. Stockholm, Laurentii Salvii, Vol. 1.

Lott, D. F. 1991. *Intraspecific Variation in the Social Systems of Wild Vertebrates.* Cambridge Studies in Behavioural Biology. New York: Cambridge University Press.

Lovejoy, C. O. 1981. The Origin of Man. *Science,* 211:341–50.

Lovejoy, O. 1978. A Biomechanical Review of the Locomotor Diversity of Early Hominids. In *Early Hominids of Africa,* Clifford Jolly (ed.). New York: St. Martin's Press.

Loy, J. 1987. The Sexual Behavior of African Monkeys and the Question of Estrus. In *Comparative Behavior of African Monkey,* E. Zucker (ed.), pp. 175–95. New York: Alan R. Liss.

Loy, T. H. 1986. Recent Advances in Blood Residue Analysis. In *Proceedings of the 24th International Archaeometry Symposium,* J. Olin and J. Blackman (eds.), pp. 54–65. Washington, DC: Smithsonian Institution Press.

Luo, Z. and D. R. Ketten. 1991. CT Scanning and Computerized Reconstructions of the Inner Ear of Multituberculate Mammals. *Journal of Vertebrate Paleontology,* 11:220–28.

McBride, L. R. 1972. *The Kahuna: Versatile Mystics of Old Hawaii.* Hilo, HI: Petroglyph Press.

McClintock, M. K. 1987. A Functional Approach to the Behavioral Endocrinology of Rodents. in *Psychology of Reproductive Behavior: An Evolutionary Perspective,* D. Crews (ed.), pp. 176–203. Englewood Cliffs, NJ: Prentice-Hall.

McCormick, B. H., T. A. DeFanti and M. D. Brown. 1987. *Visualization in Scientific Computing.* Baltimore: ACM.

McDougall, I. and T. M. Harrison. 1988. *Geochronology and Thermo-chronology by the 40Ar/39Ar Method.* Oxford Monographs on Geology and Geophysics No. 9. New York: Oxford University Press.

McGowan, C. 1989. Computed Tomography Reveals Further Details of Excalibosaurus, a Putative Ancestor for the Swordfish-Like Ichthyosaur Eurhinosaurus. *Journal of Vertebrate Paleontology,* 9:269–81.

McGuire, M., M. J. Raleigh and C. Johnson. 1983. Social Dominance in Adult Male Vervet Monkeys II; Behavior-Biochemical Relationships. *Social Science Information*, 22:311–28.

McHenry, H. M. 1986. The First Bipeds: A Comparison of the *A. afarensis* and *A. africanus* Postcranium and Implications for the Evolution of Bipedalism. *Journal of Human Evolution*, 15:177–91.

MacLarnon, A. M. 1989. Applications of the Reflex Instruments in Quantitative Morphology. *Folia Primatologica*, 53:33–49.

MacLean, P. D. 1970. The Triune Brain, Emotion, and Scientific Bias. In *The Neurosciences: Second Study Program*, F. Schmitt (ed.), pp. 336–49. New York: The Rockefeller University Press.

MacLeod, N. 1990. Digital Images and Automated Image Analysis Systems. In *Proceedings of the Michigan Morphometrics Workshop*, F. J. Rohlf and F. L. Bookstein (eds.), pp. 21–36. Ann Arbor: University of Michigan Museum of Zoology.

Madsen, D. 1985. A Biochemical Property Relating to Power Seeking in Humans. *American Political Science Review*, 79:448–57.

Mann, A. 1973. Australopithecine Age at Death. *Transvaal Museum Bulletin*, 14:11.

Marsh, J. L. and M. W. Vannier. 1983. The 'Third' Dimension in Craniofacial Surgery. *Plastic and Reconstructive Surgery*, 71:759.

Martau, P. A., H. G. Caine and D. K. Candland. 1985. Reliability of the Emotions Profile Index, Primate Form, with *Papio hamadryas, Macaca fuscata*, and Two *Saimiri* Species. *Primates*, 26(4): 501–5.

Marzke, M. W. 1983. Joint Functions and Grips of the *Australopithecus afarensis* Hand, with Special Reference to the Region of the Capitate. *Journal of Human Evolution*, 12:197–211.

Mayr, E. 1992. *One Long Argument: Charles Darwin and the Genesis of Modern Evolutionary Thought*. Cambridge: Harvard University Press.

_____. 1976. *Evolution and the Diversity of Life: Selected Essays*. Cambridge: Belknap Press of Harvard University.

Mellars, P. and C. Stringer (eds.). 1989. *The Human Revolution: Behavioural and Biological Perspectives on the Origins of Modern Humans*. Edinburgh: Edinburgh University Press.

Mendoza, S. P. 1984. The Psychobiology of Social Relationships. In *Social Cohesion: Essays Toward a Sociophysiological Perspective*, P. R. Barachas and S. P. Mendoza (eds.), pp. 3–29. Westport, CT: Greenwood Press.

Mendoza, S. P., D. M. Lyons and W. Saltzman. 1991. Sociophysiology of Squirrel Monkeys. *American Journal of Primatology*, 23:37–54.

Meyers, R. E. 1978. Comparative Neurology of Vocalization and Speech: Proof of a Dichotomy. In *Human Evolution: Biosocial Perspectives*, S. L. Washburn and E. R. McCown (eds.), pp. 59–73. Menlo Park, CA: Benjamin Cummings Press.

Mills, J. R. E. 1955. Ideal Dental Occlusion in the Primates. *Dental Practice*, 6:47–61.

Mitchell, C. L., S. Boinski and C. P. van Schaik. 1991. Competitive Regimes and Female Bonding in Two Species of Squirrel Monkeys (*Saimiri oerstedi* and *S. sciureus*). *Behavioral Ecology and Sociobiology*, 28:55–60.

Mohnot, S. M. 1984. Some Observations on All-Male Bands of the Hanuman Langur, *Presbytis entellus*. In *Current Primate Researches*, M. L. Roonwal, S. M. Mohnot and N. S. Rathore (eds.), pp. 343–56. Jodhpur, India: University of Jodhpur.

Money, J. 1991. The Development of Sexuality and Eroticism in Humankind. In *Heterotypical Behaviour in Man and Animals*, M. Haug, P. F. Brain and C. Aron (eds.), pp. 127–66. New York: Chapman & Hall.

Moore, J. J. 1985. Demography and Sociality in Primates. Ph.D. Thesis, Harvard University.

Moore, M. C. and C. A. Marler. 1988. Hormones, Behavior, and the Environment: An Evolutionary Perspective. In *Processing of Environmental Information in Vertebrates*, M. H. Stetson (ed.), pp. 71–83. New York: Springer-Verlag.

Moore, R. 1953. *Man, Time, and Fossils: The Story of Evolution*. New York: Alfred A. Knopf.

Newton, P. N. 1987. The Social Organization of Forest Hanuman Langurs (*Presbytis entellus*). *International Journal of Primatology*, 8:199–232.

Ney, D., E. K. Fishman and D. Magrid. 1990. Three-Dimensional Imaging of Computed Tomography: Techniques and Applications. *Proceedings of the First Conference on Visualization in Biomedical Computing*, pp. 498–506. Atlanta, May 22–25.

Oakley, K. 1961. On Man's Use of Fire, with Comments on Toolmaking and Hunting. In *The Social Life of Early Man*, S. L. Washburn (ed.), pp. 176–93. Viking Fund Publications in Anthropology, No. 31. Chicago: Aldine Publishing Company.

Oxnard, C. 1973. *Form and Pattern in Human Evolution*. Chicago: University of Chicago Press.

Oyama, S. 1985. The Ontogeny of Information: Developmental Systems and Evolution. Cambridge: Cambridge University Press.

Paabo, S., R. Higuchi and A. C. Wilson. 1989. Ancient DNA and the Polymerase Chain Reaction. *Journal of Biology and Chemistry*, 264(17): 9,709–12.

Parkin, A., H. Norwood, M. A. Erdentug and A. J. Hall. 1990. Optical Disk Archiving Using a Personal Computer: A Solution to Image Storage Problems in Diagnostic Imaging Departments. *Journal of Medical Engineering and Technology*, 14:55–59.

————. 1989. A Flexible Image Archiving System Using a Personal Computer and Optical Disk. *British Journal of Radiology*, 62:620–22.

Partridge, T. 1986. Paleoecology of the Pliocene and Lower Pleistocene Hominids of Southern Africa: How Good Is the Chronological and Paleoenvironmental Evidence? *South African Journal of Science*, 82:82–83.

Pereira, M. E. 1992. The Development of Dominance Relations Before Puberty in Cercopithecine Societies. In *Aggression and Peacefulness in Humans and Other Primates*, J. Silverberg and P. Gray (eds.), pp. 117–49. New York: Oxford University Press.

Perutz, M. 1983. Species Adaptation in a Protein Molecule. *Molecular Biology and Evolution*, 1(1): 1–28.

Peters, C. R. 1982. Electron-Optical Microscopic Study of Incipient Dental Microdamage from Experimental Seed and Bone Crushing. *American Journal of Physical Anthropology*, 57:283–301.

Petrocchi, C. 1952. Paleontologia di Sahabi (Cirenaica). I. Notizie Generale sul Giacimento Fossilifero di Sahabi: Storia Deglia Scavi i Resultati. *Rendiconti Accademia Nazionale dei Quaranta, Rome*, ser. 4, 3:9–33.

Pickford, M. 1990. Uplift of the Roof of Africa and Its Bearing on the Evolution of Mankind. *Human Evolution*, 5:1–20.

Pickford, M., B. Senut, I. Ssemmanda, D. Elepu and P. Obwona. 1988. Premiers Résultats de la Mission de l-Uganda Palaeontology Expedition á Nkondo (Pliocène du Bassin du Lac Albert, Ouganda). *Compte Rendus de l'Academie des Sciences, Paris*, Ser. 2, 306:315–20.

Pilbeam, D. 1984. The Descent of Hominoids and Hominids. *Scientific American*, 250(3): 84–96.

_____. 1972. Evolutionary Changes in the Hominoid Dentition Through Geological Time. In *Calibration of Hominoid Evolution: Recent Advances in Isotopic and Other Dating Methods Applicable to the Origin of Man*, W. W. Bishop and J. A. Miller (eds.), pp. 369–80. Edinburgh: Scottish Academic Press.

Ploog, D. 1970. Social Communication Among Animals. In *The Neurosciences: Second Study Program*, F. Schmitt (ed.), pp. 349–61. New York: The Rockefeller University Press.

Poirier, F. 1987. *Understanding Human Evolution*. Englewood Cliffs, NJ: Prentice-Hall.

Pope, A. 1967 (original 1733). Material appearing in *The Works of Alexander Pope*, Vols. 1–4, W. Elwin and J. W. Croker (eds.). New York: Gordian Press.

Potts, R. B. 1988. *Early Hominid Activities at Olduvai*. New York: Aldine de Gruyter.

_____. 1986. Temporal Span of Bone Accumulations at Olduvai Gorge and Implications for Early Hominid Foraging Behaviour. *Paleobiology*, 12:25–31.

Potts, R. B. and P. Shipman. 1981. Cutmarks Made by Stone Tools on Bones from Olduvai Gorge, Tanzania. *Nature*, 291:577–80.

Prentice, A. and R. G. Whitehead. 1987. The Energetics of Human Reproduction. *Symposium of the Zoological Society of London*, 57:275–304.

Puech, P. F. 1981. Tooth Wear in La Ferrassie Man. *Current Anthropology*, 22:424–25.

_____. 1979. The Diet of Early Man: Evidence from Abrasion of Teeth and Tools. *Current Anthropology*, 20:590–92.

Puech, P. F., H. Albertini and C. Serratrice. 1983. Tooth Microwear and Dietary Patterns in Early Hominids from Laetolil, Hadar and Olduvai. *Journal of Human Evolution*, 12:721–29.

Rajpurohit, L. S. 1991. Resident Male Replacement, Formation of a New Male Band, and Paternal Behavior in *Presbytis entellus*. *Folia Primatologica*, 57:159–64.

Raleigh, M. J., G. L. Brammer, M. T. McGuire, A. Yuwiler, E. Geller and C. K. Johnson. 1983. Social Status Related Differences in the Behavioral Effects of Drugs in Vervet Monkeys (*Cercopithecus aethiops sabaeus*).

In *Hormones Drugs and Social Behavior in Primates*, H. D. Steklis and A. S. Kling (eds.), pp. 83–106. New York: Spectrum Publications.

Raleigh, M. J., M. T. McGuire, G. L. Brammer and A. Yuwiler. 1984. Social and Environmental Influences on Blood Serotonin Concentrations in Monkeys. *Archives of General Psychiatry*, 41:405–410.

Reilly, S. 1990. Comparative Ontogeny of Cranial Shape in Salamanders Using Resistant Fit Theta Rho Analysis. In *Proceedings of the Michigan Morphometrics Workshop*, F. J. Rohlf and F. L. Bookstein (eds.), pp. 311–22. Ann Arbor: University of Michigan Museum of Paleontology.

Rekow, E. D. 1989. CAD/CAM for Crowns. *Transactions of the Academy of Dental Materials: Proceedings of a Conference on CAD/CAM in Dentistry*, 58:17–20.

_____. 1987. Computer-Aided Design and Manufacturing in Dentistry: A Review of the State of the Art. *Journal of Prosthetic Dentistry*, 58:512–16.

Rekow, E. D. and A. G. Erdman. 1985. Comparison of Techniques for Acquiring High-Resolution Three-Dimensional Computer-Based Data Directly from the Human Mouth. *Advances in Bioengineering, American Society of Mechanical Engineers*, 11:139–40.

Restak, R. M. 1984. *The Brain*. Toronto: Bantam Books.

Rhinelander, P. H. 1973. *Is Man Incomprehensible to Man?* Stanford: Stanford Alumni Association.

Richard, A. F. 1985. *Primates in Nature*. New York: W. H. Freeman.

_____. 1981. Changing Assumptions in Primate Ecology. *American Anthropologist*, 83(198): 517–33.

Richtsmeier, J. T. 1990. Quantitative Analysis of Three-Dimensional Data. *Society of Photo-Optical Instrumentation Engineers, Proceedings*, 1,380:49.

Robinson, J. T. 1969. Dentition and Adaptation in Early Hominids. *Proceedings of the 8th International Congress in Anthropological and Ethnological Science I*: Anthropology, Tokyo, 1968, pp. 302–5.

Robinson, J. T. 1963. Adaptive Radiation of the Australopithecines and the Origin of Man. In *African Ecology and Human Evolution*, F. C. Howell and F. Bourliere (eds.). Chicago: Aldine.

_____. 1953. Meganthropus, Australopithecines, and Hominids. *American Journal of Physical Anthropology*, 11:1–38.

Rodman, P. S. 1988. Resources and Group Size of Primates. In *The Ecology of Social Behavior*, C. N. Slobodchikoff (ed.), pp. 83–108. New York: Academic Press.

Rohlf, F. J. 1990. An Overview of Image Processing Techniques and Analysis for Morphometrics. in *Proceedings of the Michigan Morphometric Workshop*, F. J. Rohlf and F. L. Bookstein (eds.), pp. 37–60. Ann Arbor: University of Michigan Museum of Zoology.

Romer, A. S. 1964. *The Vertebrate Story*. Chicago: University of Chicago Press.

Rowell, T. E. 1979. How Would We Know if Social Organization Were *Not* Adaptive? In *Primate Ecology and Human Origins: Ecological Influences on Social Organization*, I. S. Bernstein and E. O. Smith (eds.), pp. 1–22. New York: Garland.

Rowell, T. E. and J. Chism. 1986. Sexual Dimorphism and Mating Systems: Jumping to Conclusions. *Human Evolution*, 1(3): 215–19.

Ruff, C. B. and F. P. Leo. 1986. Use of Computed Tomography in Skeletal Structure Research. *Yearbook of Physical Anthropology*, 29:181–96.

Ruvolo, M., T. Disotell, M. Allards, W. Brown and R. Honeycutt. 1991. Resolution of the African Hominoid Trichotomy by Use of a Mitochondrial Gene Sequence. *Proceedings, National Academy of Sciences USA*, 88:1,570–74.

Ryan, A. S. 1979. Wear Striation Direction on Primate Teeth: A Scanning Electron Microscope Examination. *American Journal of Physical Anthropology*, 50:155–68.

Sacher, G. A. and E. F. Staffeldt. 1974. Relation Between Gestation Time and Brain Weight of Placental Mammals: Implications for the Theory of Vertebrate Growth. *American Naturalist*, 105:593–615.

Sadler, L. L., P. Neumann and X. Chen. In press. A Method for the Analysis of Three Dimensional Shape Change. *Proceedings of the National Computer Graphics Association*.

Sapolsky, R. M. 1991. Testicular Function, Social Rank and Personality Among Wild Baboons. *Psychoneuroendocrinology*, 16(4): 281–93.

Sarich, V. 1969. Pinniped Origins and the Rate of Evolution of Carnivore Albumins. *Systematic Zoology*, 18:286–95.

Sarich, V. M. and A. C. Wilson. 1967. Immunological Time Scale for Hominid Evolution. *Science*, 158:1,200–1,203.

Sawaguchi, T. 1990. Relative Brain Size, Stratification, and Social Structure in Anthropoids. *Primates*, 31(2): 257–72.

Sawaguchi, T. and H. Kudo. 1990. Neocortical Development and Social Structure in Primates. *Primates*, 31(2): 238–89.

Schaller, G. B. 1972. *The Serengeti Lion: A Study of Predator-Prey Relations*. Chicago: The University of Chicago Press.

Schaller, G. B. and G. R. Lowther. 1969. The Relevance of Carnivore Behavior to the Study of Early Hominids. *Southwestern Journal of Anthropology*, 25:307–41.

Schick, K. D. 1986. Stone Age Sites in the Making. *B.A.R. International Series*, 319.

Schmid, P. 1983. Eine Rekonstruktion des Skelettes von A.L. 288–1 (Hadar) und deren Konsequenzen. *Folia Primatologica*, 40:283–306.

Schneider, M. L., C. F. Moore, S. J. Suomi and M. Champoux. 1991. Laboratory Assessment of Temperament and Environmental Enrichment in Rhesus Monkey Infants (*Macaca mulatta*). *American Journal of Primatology*, 25:137–55.

Schultz, A. H. 1971. The Rise of Primatology in the Twentieth Century. *Proceedings of the 3rd International Congress of Primatology*, 1:2–15.

Schurr, T. G., S. W. Ballinger, Y-Y. Gan, J. A. Hodge, D. A. Merriwether, D. N. Lawrence, W. Knowler, K. M. Weiss and D. C. Wallace. 1990. Amerindian Mitochondrial DNAs Have Rare Asian Mutations at High Frequencies, Suggesting They Derived From Four Primary Maternal Lineages. *American Journal of Human Genetics*, 46:613–23.

Schwarcz, H. P., R. Grun, B. Vandermeersch, O. Bar-Yosef, H. Valladas and E. Tchernov. 1988. ESR Dates for the Hominid Burial Site of Qafzeh in Israel. *Journal of Human Evolution*, 17:733–37.

Scott, P. J. 1985. Direct Three-Dimensional Measurements with the Reflex Instruments. *Society of Photo-Optical Instrumentation Engineers Proceedings*, 602:31–33.

———. 1981. The Reflex Plotters: Measurement without Photographs. *Photogrammetric Record*, 24:435–36.

Selvik, G. 1990. Roentgen Stereophotogrammetric Analysis. *Acta Radiologica*, 31:113–26.

Setchell, D. J. 1984. The Reflex Microscope—An Assessment of the Accuracy of 3-Dimensional Measurements Using a new Meteorological Instrument. *Journal of Dental Research*, 63:493.

Shipman, P. L. 1983. Early Hominid Lifestyle. Hunting and Gathering or Foraging and Scavenging? Paper presented at the 52nd Annual Meeting, American Association of Physical Anthropologists, Indianapolis, April, 1983.

Shipman, P., G. Foster and M. Schoeninger. 1984. Burnt Bones and Teeth: An Experimental Study of Color Morphology, Crystal Structure and Shrinkage. *Journal of Archaeological Science*, 11:307–25.

Shively, C. A., G. L. Brammer, J. R. Kaplan, M. J. Raleigh and S. B. Manuck. 1991. The Complex Relationship Between Behavioral Attributes, Social Status, and Whole Blood Serotonin in Male *Macaca fascicularis*. *American Journal of Primatology*, 23:99–112.

Shkurkin, G. V., A. J. Almquist, A. A. Pfeihofer and E. L. Stoddard. 1975. Scanning Electron Microscopy of Dentition: Methodology and Ultrastructural Morphology of Tooth Wear. *Journal of Dental Research*, 54:402–6.

Short, R. V. 1987. The Biological Basis for the Contraceptive Effects of Breast Feeding. *International Journal of Gynecology and Obstetrics (Supplement)*, 25:207–15.

Sillen, A. and C. K. Brain. 1988. Evidence from the Swartkrans Cave for the Earliest Use of Fire. *Nature*, 336:446–66.

Simons, E. L. 1989. Human Origins, *Science*, 245:1,343–50.

Simpson, G. G. 1964. *This View of Life: The World of an Evolutionist*. New York: Harcourt, Brace and World.

Smuts, B. 1992. Male Aggression Against Women: An Evolutionary Perspective. *Human Nature*, 3(1).

Smuts, B. B., D. L. Cheney, R. M. Seyfarth, R. W. Wrangham and T. T. Struhsaker (eds.). 1987. *Primate Societies*. Chicago: University of Chicago Press.

Sommer, V. 1988. Male Competition and Coalitions in Langurs (*Presbytis entellus*) at Jodhpur, Rajasthan, India. *Human Evolution*, 3:261–78.

Sommer, V. and S. M. Mohnot. 1985. New Observations on Infanticides Among Hanuman Langurs (*Presbytis entellus*) Near Jodhpur (Rajasthan/India). *Behavioral Ecology and Sociobiology*, 16(3): 245–48.

Speculand, B., G. W. Butcher and C. D. Stephens. 1988. Three Dimensional Measurement: The Accuracy and Precision of the Reflex Microscope. *British Journal of Oral and Maxillofacial Surgery*, 26:276–83.

Springer, S. and G. Deutsch. 1981. *Left Brain, Right Brain*. San Francisco: W. H. Freeman.

Stamps, J. A. 1991. Why Evolutionary Issues Are Reviving Interest in Proximate Behavioral Mechanisms. *American Zoologist*, 31:338–48.

Steklis, H. D. and A. Kling. 1985. Neurobiology of Affiliative Behavior in Nonhuman Primates. In *The Psychobiology of Attachment and Separation*, M. Reite and T. Field (eds.), pp. 93–134. New York: Academic Press.

Steklis, H. D., M. J. Raleigh, A. S. Kling and K. Tachiki. 1986. Biochemical and Hormonal Correlates of Dominance in All-Male Groups of Squirrel Monkeys (*Saimiri sciureus*). *American Journal of Primatology*, 11:133–45.

Steklis, H. D. and A. Walter. 1991. Culture, Biology, and Human Behavior: A Mechanistic Approach. *Human Nature*, 2(2): 137–69.

Steklis, H. D. and C. H. Whiteman. 1989. Loss of Estrus in Human Evolution: Too Many Answers, Too Few Questions. *Ethology and Sociobiology*, 10:417–34.

Stevenson-Hinde, J., R. Stillwell-Barnes and M. Zunz. 1980. Subjective Assessment of Rhesus Monkeys Over Four Successive Years. *Primates*, 21(1): 66–82.

Stevenson-Hinde, J. and M. Zunz. 1978. Subjective Assessment of Individual Rhesus Monkeys. *Primates*, 19(3): 473–82.

Stewart, K. J. and A. H. Harcourt. 1987. Gorillas: Variation in Female Relationships. In *Primate Societies*, B. B. Smuts, D. L. Cheney, R. M. Seyfarth, R. W. Wrangham and T. T. Struhsaker (eds.), pp. 155–64. Chicago: The University of Chicago Press.

Stilwell, C. E. and M. R. Heath. 1987. The Repeatability of Measurements of 3-D Soft Tissue Contours Using a Reflex Microscope. *Journal of Dental Research*, 66:883.

Stini, W. A. 1985. Sexual Dimorphism and Nutrient Reserves. In *Sexual Dimorphism in* Homo sapiens, R. Hall (ed.), pp. 391–419. New York: Praeger.

Stirton, R. A. 1959. *Time, Life, and Man*. New York: John Wiley and Sons.

Straus, W., Jr. 1949. The Riddle of Man's Ancestry. *Quarterly Review of Biology*, 24:200–23.

Stringer, C. B. and P. Andrews. 1988. Genetic and Fossil Evidence for the Origin of Modern Humans. *Science*, 239:1,263–68.

Stringer, C. B., R. Grun, H. P. Schwarcz and P. Goldberg. 1989. ESR Dates for the Hominid Burial Site of Es Skhul in Israel. *Nature*, 338:756–58.

Struhsaker, T. T. 1967. Social Structure Among Vervet Monkeys. *Behaviour*, 29:83–121.

Strum, S. 1983. Baboon Cues for Eating Meat. *Journal of Human Evolution*, 12:327–36.

Sueoka, N. 1961. Correlation Between Base Composition of Deoxyribonucleic Acid and Amino Acid Composition of Protein. *Proceedings, National Academy of Sciences (USA)*, 47:1,141–49.

Sugiyama, Y. 1967. Social Organization of Hanuman Langurs. In *Social Communication Among Primates*, S. A. Altmann (ed.), pp. 221–36. Chicago: University of Chicago Press.

_____. 1965. On the Social Change of Hanuman Langurs (*Presbytis entellus*) in Their Natural Condition. *Primates*, 6:381–418.

Surbey, M. K. 1987. Anorexia Nervosa, Amenorrhea, and Adaptation. *Ethology and Sociobiology*, 8(3S): 47–62.

Susman, R. L. and N. Creel. 1979. Functional and Morphological Affinities of the Subadult Hand (OH 7) from Olduvai Gorge. *American Journal of Physical Anthropology*, 51:311–32.

Susman, R. L., J. T Stern and W. L. Jungers. 1984. Arboreality and Bipedality in the Hadar Hominids. *Folia Primatologica*, 43:113–56.

Suzuki, A. 1975. The Origin of Hominid Hunting: A Primatological Perspective. In *The Socioecology and Psychology of Primates*, R. Tuttle (ed.). The Hague, The Netherlands: Mouton.

Szalay, F. S. and R. K. Costello. 1991. Evolution of Permanent Estrus Displays in Hominids. *Journal of Human Evolution*, 20:439–64.

Taff, M. A. 1990. Social Dynamics of Langur (*Presbytis entellus*) All-Male Groups: A Captive Study. Ph.D. Thesis, University of California, Berkeley.

Taieb, M., Y. Coppen, D. Johanson and J. Kalb. 1972. Depots Sedimentaires et Faunes du Plio-Pleistocene de la Basse Vallee de l'Awash (Afar Central, Ethiopie). *Comptes Rendus de l'Academie des Sciences, Paris*, 275D:819–22.

Tanner, N. M. 1981. *On Becoming Human*. New York: Cambridge University Press.

Tate, J. R. and C. E. Cann. 1982. High-Resolution Computed Tomography for the Comparative Study of Fossil and Extant Bone. *American Journal of Physical Anthropology*, 58:67–73.

Teaford, M. F. 1990. Measurement of Teeth Using the Reflex Microscope. *Society of Photo-Optical Instrumentation Engineers Proceedings*, 1,380:49.

_____. 1988. A Review of Dental Microwear and Diet in Modern Mammals. *Scanning Microscopy*, 2:1,149–66.

_____. 1985. Molar Microwear and Diet in the Genus *Cebus*. *American Journal of Physical Anthropology*, 66:363–70.

Teaford, M. F. and K. E. Byrd. 1989. Differences in Tooth Wear as an Indicator of Changes in Jaw Movement in the Guinea Pig *Cavia Porcellus*. *Archives of Oral Biology*, 34:929–36.

Teaford, M. F. and O. J. Oyen. 1989a. Differences in the Rate of Molar Wear Between Monkeys Raised on Different Diets. *Journal of Dental Research*, 68:1,513–18.

_____. 1989b. In Vivo and In Vitro Turnover in Dental Microwear. *American Journal of Physical Anthropology*, 80:447–60.

_____. 1989c. Live Primates and Dental Replication: New Problems and New Techniques. *American Journal of Physical Anthropology*, 80:73–81.

Teaford, M. F. and A. C. Walker. 1983a. Dental Microwear in Adult and Still-Born Guinea Pigs (*Cavia porcellus*). *Archives of Oral Biology*, 28:1,077–81.

_____. 1983b. Prenatal Jaw Movements in the Guinea Pig, *Cavia Porcellus*: Evidence from Patterns of Tooth Wear. *Journal of Mammalogy*, 64:534–36.

Teleki, G. 1973. *The Predatory Behavior of Wild Chimpanzees.* Lewisburg, PA: Bucknell University Press.

Terborgh, J. T. and C. H. Janson. 1986. The Socioecology of Primate Groups. *Annual Review of Ecology and Systematics,* 17:111–36.

Thierry, B. 1985. Patterns of Agonistic Interactions in Three Species of Macaque (*Macaca mulatta, M. fascicularis, M. tonkeana*). *Aggressive Behavior,* 11:223–33.

Thomason, J. 1990. A Functional Interpretation of Bone Distribution in Mammalian Skulls. *Journal of Vertebrate Paleontology,* 10:46A.

Thompson, S. J. 1988. *A Chronology of Geological Thinking from Antiquity to 1899.* Metuchen, NJ: The Scarecrow Press.

Tierney, A. J. 1986. The Evolution of Learned and Innate Behavior: Contributions From Genetics and Neurobiology to a Theory of Behavioral Evolution. *Animal Learning & Behavior,* 14(4): 339–48.

Tobias, P. V. 1980. "*Australopithecus afarensis*" and *A. africanus*: Critique and an Alternative Hypothesis. *Palaeontologia Africana,* 23:1–17.

Toth, N. 1985. Archaeological Evidence for Preferential Right-Handedness Implications. *Journal of Human Evolution,* 14:607–14.

_____. 1982. The Stone Technologies of Early Hominids from Koobi Fora, Kenya: An Experimental Approach. Ph.D. Thesis in Anthropology, University of California, Berkeley.

Trivers, R. L. 1972. Parental Investment and Sexual Selection. In *Sexual Selection and the Descent of Man,* B. Campbell (ed.), pp. 136–79. Chicago: Aldine.

Tuppy, H. 1958. Uber die Artspeziftat der Protein-Struktur. In *Symposium on Protein Structure,* A. Neuberger (ed.), pp. 66–76. New York: John Wiley and Sons.

Tuttle, R. 1988. What's New in African Paleoanthropology? *Annual Review of Anthropology,* 17:391–426.

_____. 1969. Knuckle-Walking and the Problem of Human Origins. *Science,* 166:953–51.

Ubelaker, D. H. 1990a. Positive Identification of American Indian Skeletal Remains from Radiograph Comparison. *Journal of Forensic Science,* 35:466–81.

_____. 1990b. Importance to Forensic Science of Permanent Curation of Museum Collections of Human Remains. *Journal of Forensic Science,* 35:513.

Ubelaker, D. H. and L. G. Grant. 1989. Human Skeletal Remains: Preservation or Reburial? *Yearbook of Physical Anthropology,* 32:249–87.

Valladas, H., J. L. Joron, G. Valladas, O. Bar-Yosef, A. Belfer-Cohen, P. Goldberg, H. Laville, L. Meignen, Y. Rak, E. Tchernov, A. M. Tillier and B. Vandermeersch. 1987. Thermoluminescence Dates for the Neanderthal Burial Site at Kebara in Israel. *Nature,* 330:159–60.

Valladas, H., J. L. Reyss, J. L. Joron, G. Valladas, O. Bar-Yosef and B. Vandermeersch. 1988. Thermoluminescence Dating of Mousterian "Proto-Cro-Magnon" Remains from Israel and the Origin of Modern Man. *Nature,* 331:614–16.

Van Der Bijl, P., J. De Waal, A. Botha and W. P. Dreyer. 1989. Assessment of Occlusal Tooth Wear in Vervet Monkeys by Reflex Microscopy. *Archives of Oral Biology*, 34:723–29.

Vandermeersch, B. 1989. The Evolution of Modern Humans: Recent Evidence from Southwest Asia. In *The Human Revolution*, P. Mellars and C. Stringer (eds.), pp. 155–64. Edinburgh: Edinburgh University Press.

Vannier, M. W. and G. C. Conroy. 1989a. Imaging Workstations for Computer-Aided Primatology: Promises and Pitfalls. *Folia Primatologica*, 53:7–21.

_____. 1989b. Three-Dimensional Surface Reconstruction Software System for IBM Personal Computers. *Folia Primatologica*, 53:22–32.

Vannier, M. W., G. C. Conroy, J. L. Marsh and R. H. Knapp. 1985. Three-Dimensional Cranial Surface Reconstructions Using High-Resolution Computed Tomography. *American Journal of Physical Anthropology*, 67:299–311.

Van Poppel, B. M., A. R. Bakker and J. B. M. Wilmink. 1990. A Package for Cost and Critical Analysis of Picture Archiving and Communication Indicating its True Yield. *Medical Information*, 15:67–75.

van Schaik, C. P. 1989. The Ecology of Social Relationships Amongst Female Primates. In *Comparative Socioecology: The Behavioural Ecology of Humans and Other Mammals*, V. Standen and R. A. Foley (eds.), pp. 195–218. Boston: Blackwell Scientific Publications.

_____. 1983. Why Are Diurnal Primates Living in Groups? *Behaviour*, 87:120–43.

van Schaik, C. P. and R. I. M. Dunbar. 1990. The Evolution of Monogamy in Large Primates: A New Hypothesis and Some Crucial Tests. *Behaviour*, 115(1–2): 30–61.

van Schaik, C. P. and J. A. R. A. M. van Hooff. 1983. On the Ultimate Causes of Primate Social Systems. *Behaviour*, 85:91–117.

van Schaik, C. P. and M. A. van Noordwijk. 1985. Evolutionary Effects of the Absence of Felids on the Social Organization of the Macaques on the Island of Simeulue (*Macaca fascicularis fusca*, Miller 1903). *Folia Primatologica*, 44:138–47.

Vigilant, L., R. Pennington, H. Harpending, T. D. Kocher and A. C. Wilson. 1989. Mitochondrial DNA Sequences in Single Hairs From a Southern African Population. *Proceedings, National Academy of Sciences (USA)*, 86:9,350–54.

Vigilant, L., M. Stoneking, H. Harpending, K. Hawkes and A. C. Wilson. 1991. African Populations and the Evolution of Mitochondrial DNA. *Science*, 253:1,503–7.

von Koenigswald, G. H. R. 1955. *Begegnungen mit dem Vormenschen*. Düsseldorf, W. Germany: Diederichs.

Vrba, E. S. 1985. Environment and Evolution: Alternative Causes of the Temporal Distribution of Evolutionary Events. *South African Journal of Science*, 81:263–66.

_____. 1982. Biostatigraphy and Chronology, Based Particularly on Bovidae, of Southern African Hominid-Associated Assemblages,

Makapansgat, Sterkfontein, Taung, Kromdraai, Swartkrans, also Elandsfontein (Saldanha), Broken Hill (now Kabwe) and Cave of Hearths. In *Prelirage, ler Congr. Internat. Paleont. Humaine. Nice: Cent. Nat. Rech. Sci.* (Vol. 2), Henry De Lumley and Marie-Antoinette De Lumley (eds.), pp. 707–52.

_____. 1980. Evolution, Species and Fossils: How Does Life Evolve? *South African Journal of Science*, 76:61–84.

_____. 1975. Some Evidence of Chronology and Palaeoecology of Sterkfontein, Swartkrans and Kromdraai from the Fossil Bovidae. *Nature*, 254:301–4.

Walker, A. C. 1981. Diet and Teeth—Dietary Hypotheses and Human Evolution. *Philosophical Transactions of the Royal Society, London*, 292(B): 57–64.

Walker, A. C., H. N. Hoeck and L. Perez. 1978. Microwear of Mammalian Teeth as an Indicator of Diet. *Science*, 201:908–10.

Walker, A. C., R. E. F. Leakey, J. M. Harris and F. H. Brown. 1986. 2.5 myr Australopithecus Boisei from West of Lake Turkana, Kenya. *Nature*, 322:517–22.

Walker, A. and M. Pickford. 1983. New Postcranial Fossils of *Proconsul africanus* and *Proconsul nyanzae*. In *New Interpretations of Ape and Human Ancestry*, R. Ciochon and R. Corruccini (eds.), pp. 325–51. New York: Plenum.

Walker, A. C. and M. F. Teaford. 1989. Inferences from Quantitative Analysis of Dental Microwear. *Folia Primatologica*, 53:177–89.

Walker, P. L. 1976. Wear Striations of the Incisors of Cercopithecoid Monkeys as an Index of Diet and Habitat Preference. *American Journal of Physical Anthropology*, 45:299–308.

Walter, R. C., P. C. Manega, R. E. Drake and G. H. Curtis. 1991. Laserfusion 40Ar/39Ar Dating of Bed 1, Olduvai Gorge, Tanzania. *Nature*, 354:145–49.

Walters, J. R. and R. M. Seyfarth. 1987. Conflict and Cooperation. In *Primate Societies*, B. B. Smuts, D. L. Cheney, R. M. Seyfarth, R. W. Wrangham and T. T. Struhsaker (eds.), pp. 306–17. Chicago: University of Chicago Press.

Ward, R. H., B. L. Frazier, K. Dew-Jager and S. Paabo. 1991. Extensive Mitochondrial Diversity With a Single Amerindian Tribe. *Proceedings of the National Academy of Sciences (USA)*, 88:8,720–24.

Washburn, S. L. 1982. Language and the Fossil Record. *Anthropology UCLA*, 7:231–38.

_____. 1975. Ecology and Australopithecine Taxonomy. *American Anthropologist*, 77:618.

_____. 1973a. The Promise of Primatology. *American Journal of Anthropology*, 38(2): 177–82.

_____. 1973b. Primate Studies and Human Evolution. In *Nonhuman Primates and Medical Research*, G. H. Bourne (ed.), pp. 467–85. New York: Academic Press.

_____. 1973c. The Evolution Game. *Journal of Human Evolution*, 2:557–61.

_____. 1951. The New Physical Anthropology. *Transactions of the New York Academy of Sciences*, Series II, 13(7): 298–304.

Washburn, S. L. and I. DeVore. 1961. Social Behavior of Baboons and Early Man. In *Social Life of Early Man*, S. L. Washburn (ed.), pp. 91–103. Chicago: Aldine Press.

Washburn, S. L. and D. Hamburg. 1965. The Implications of Primate Research. In *Primate Behavior*, I. DeVore (ed.), pp. 607–22. New York: Holt, Rinehart, and Winston.

Washburn, S. L. and R. Moore. 1960. *Ape into Human*. Boston: Little, Brown.

Wasser, S. K. and D. P. Barash. 1983. Reproductive Suppression Among Female Mammals: Implications for Biomedicine and Sexual Selection Theory. *Quarterly Review of Biology*, 58:513–38.

Wasser, S. K. and D. Y. Isenberg. 1986. Reproductive Failure Among Women: Pathology or Adaptation. *Journal of Psychosomatic Obstetrics and Gynaecology*, 5:153–75.

Watson, J. D. 1968. *The Double Helix*. New York: Atheneum Press.

Watson, J. D. and F. H. C. Crick. 1953. A Structure for Deoxyribose Nucleic Acid. *Nature*, 171:737–38.

Watts, D. P. 1991. Strategies of Habitat Use by Mountain Gorillas. *Folia Primatologica*, 56:1–16.

_____. 1989. Infanticide in Mountain Gorillas: New Cases and Reconsideration of the Evidence. *Ethology*, 81:1–18.

_____. 1985. Relations Between Group Size and Composition and Feeding Competition in Mountain Gorilla Groups. *Animal Behaviour*, 33:72–85.

_____. 1984. Composition and Variability of Mountain Gorilla Diets in the Central Virungas. *American Journal of Primatology*, 7:323–56.

Wegener, A. 1924. *The Origin of Continents and Oceans*, translated by J. G. A. Skerl. London: Methuen and Company.

West-Eberhard, M. J. 1987. Flexible Strategy and Social Evolution. In *Animal Societies: Theories and Facts*, Y. Ito, J. L. Brown and J. Kikawa (eds.), pp. 35–51. Tokyo: Japan Sci. Soc. Press.

White, T. D. 1985. Acheulian Man in Ethiopia's Middle Awash Valley: The Implications of Cutmarks on the Bodo Cranium. 8th Kroon Lecture, Stichting Nederlands Museum voor Anthropologie en Praehistorie, Amsterdam. J. Enschede en zonen, Haarlem.

_____. 1984. Pliocene Hominids from the Middle Awash. *Courier Forschungeinstitut Senckenberg*, 69:57–68.

_____. 1981. Primitive Hominid Canine from Tanzania. *Science*, 213:348.

_____. 1980. Evolutionary Implications of Pliocene Hominid Footprints. *Science*, 208:175–76.

White, T. D., D. Johanson and W. Kimbel. 1981. *Australopithecus africanus*: Its Phyletic Position Reconsidered. *South African Journal of Science*, 77:445–70.

Wicander, R. and J. S. Monroe. 1989. *Historical Geology: Evolution of the Earth and Life Through Time*. St. Paul: West.

Williams, G. C. 1966. *Adaptation and Natural Selection*. Princeton: Princeton University Press.

Williamson, P. G. 1985. Evidence for an Early Plio-Pleistocene Rainforest Expansion in East Africa. *Nature*, 314:931.

Wilson, E. O. 1975. *Sociobiology: The New Synthesis*. Cambridge: Harvard University Press.

Wilson, E. O. and C. J. Lumsden. 1991. Holism and Reduction in Sociobiology: Lessons from the Ants and Human Culture. *Biology and Philosophy*, 6:401–12.

Wilson, J. T. 1976. *Continents Adrift and Continents Aground: Readings from Scientific American with Introductions by J. Tuzo Wilson.* San Francisco: W. H. Freeman.

Wingfield, J. C. 1984. Androgens and Mating Systems: Testosterone-Induced Polygyny in Normally Monogamous Birds. *Auk*, 101:665–71.

Wingfield, J. C., R. E. Hegner, A. M. Dufty, Jr. and G. F. Ball. 1990. The "Challenge Hypothesis": Theoretical Implications for Patterns of Testosterone Secretion, Mating Systems, and Breeding Strategies. *American Naturalist*, 136(6): 829–46.

Wolpoff, M. H. 1980. *Paleoanthropology.* New York: Knopf.

Wrangham, R. W. 1987. Evolution of Social Structure. In *Primate Societies*, B. B. Smuts, D. L. Cheney, R. M. Seyfarth, R. W. Wrangham and T. T. Struhsaker (eds.), pp. 282–96. Chicago: The University of Chicago Press.

_____. 1986. Ecology and Social Relationships in Two Species of Chimpanzee. In *Ecological Aspects of Social Evolution: Birds and Mammals*, D. I. Rubenstein and R. W. Wrangham (eds.), pp. 352–78. Princeton: Princeton University Press.

_____. 1980. An Ecological Model of Female-Bonded Primate Groups. *Behaviour*, 75:262–300.

_____. 1979. On the Evolution of Ape Social Systems. *Social Science Information*, 18:335–68.

Wright, R. S. V. 1972. Imitative Learning of a Flaked Stone Technology—The Case of An Orangutan. *Mankind*, 8(4): 296–306.

Yasuda, T., Y. Hashimoto, S. Yokoi and J-I. Toriwaki. 1990. Computer System for Craniofacial Surgical Planning Based on CT Images. *Institute of Electrical and Electronic Engineers, Transactions on Medical Imaging*, 9:270–80.

Young, J. M. and B. R. Altschuler. 1981. Topographic Mapping of Oral Structures, Problems, and Applications in Prosthodontics. *Society of Photo-Optical Instrumentation Engineers, Proceedings*, 283:70–77.

Zuckerkandl, E. 1965. Evolutionary Divergence and Convergence in Proteins. In *Evolving Genes and Proteins*, Vernon Bryson and Henry Vogel (eds.), pp. 97–166. New York: Academic Press.